Columbia University

Contributions to Education

Teachers College Series

No. 706

AMS PRESS
NEW YORK

THE MATHEMATICS IN CERTAIN ELEMENTARY SOCIAL STUDIES IN SECONDARY SCHOOLS AND COLLEGES

By Eugene W. Hellmich, Ph.D.

TEACHERS COLLEGE, COLUMBIA UNIVERSITY
CONTRIBUTIONS TO EDUCATION, 706

Published with the Approval of
Professor William D. Reeve, Sponsor

BUREAU OF PUBLICATIONS
Teachers College, Columbia University
NEW YORK CITY
1937

Library of Congress Cataloging in Publication Data

Hellmich, Eugene William, 1902-
The mathematics in certain elementary social studies in secondary schools and colleges.

Reprint of the 1937 ed., issued in series: Teachers College, Columbia University. Contributions to education, 706.

Originally presented as the author's thesis, Columbia.

Bibliography: p.

1. Mathematics--Study and teaching. 2. Social sciences--Study and teaching (Elementary) 3. Economics, Mathematical. 4. Statistics--Graphic methods. I. Title II. Series: Columbia University. Teachers College. Contributions to education, no. 706.

QA11.H4 1972 510'.7'1073 78-176854
ISBN 0-404-55706-6

Reprinted by Special Arrangement with Teachers College Press, New York, New York

From the edition of 1937, New York
First AMS edition published in 1972
Manufactured in the United States

AMS PRESS, INC.
NEW YORK, N. Y. 10003

ACKNOWLEDGMENTS

The writer takes pleasure in expressing his deep and sincere gratitude to his sponsor, Professor William D. Reeve, at whose suggestion the study was undertaken, and whose aid and encouragement have been invaluable; to his advisers, Professor John R. Clark and Professor Harold F. Clark, for their helpful advice and suggestions throughout the course of the study; and to Professor Clifford B. Upton, who generously gave of his time and thought.

E. W. H.

CONTENTS

THE MATHEMATICS IN CERTAIN ELEMENTARY SOCIAL STUDIES IN SECONDARY SCHOOLS AND COLLEGES

CHAPTER I

PROBLEM AND PROCEDURE

INTRODUCTION

MODERN secondary education is a vast enterprise engaged in assisting individuals to fit themselves into a complex society where they may live contentedly and to which they may contribute effectively. The responsibility for accomplishing this education is entrusted to a secondary school organization that has been made intricate by the rapid increase in the number of secondary school pupils and by the constant growth in the knowledge of the individual and of society. One of the primary functions of every secondary school is the selection of the materials of instruction to meet the demands of the increased number of pupils and of a changing society, and to adapt these materials to the varying needs and abilities of the pupils in order to carry out the fundamental purpose of secondary education.

The various changes in the social order and the new developments in psychology exert a potent influence in the choice of subject matter and in the emphasis ascribed to certain divisions of that subject matter in secondary education. The general lack of understanding of even the simpler problems involved in the recent economic crisis and the realization that most of the individual modes of thinking and forms of behavior are to be explained as manifestations of group tendencies and achievements have added vigor to the demand for a larger place in secondary education for those materials of instruction that purpose to make clear the development and the organization of society and the function of individuals and groups of individuals within that society.

To the end that secondary school pupils may be made conversant with modern economic and political conditions, an increased emphasis has been directed toward the social studies in secondary education.[1]

DEFINITION OF THE SOCIAL STUDIES

The social studies have been defined as those subjects whose subject matter relates directly to the organization and the development of society, and to man as a member of social groups.[2] More specifically, the social studies, or the social sciences, as some prefer, is the term used somewhat arbitrarily in this study to describe the content embraced in the conventional materials of history, geography, civics, economics, and sociology.[3]

The more or less arbitrary selection of a definition was necessitated by the great variation in the meanings that are attached to the social studies. For example, the Commission on the Social Studies includes anthropology and psychology among these subjects.[4] Kimmel defines the social studies in secondary schools as comprising a group of subjects including history, civics, economics, and sociology, with certain relatively undefined boundaries and ramifications in the subject matter of geography, vocations, ethics, and homemaking.[5] Definitions are found also to include within the province of the social studies literature and esthetics,[6] biology,[7] and frequently, at the elementary school level, hygiene and health, and the natural sciences.

[1] See, for example:

Beard, Charles A., *A Charter for the Social Sciences in the Schools.* 1932.

Counts, George S., *The Social Foundations of Education.* 1934.

Dawson, Edgar and Others, *Teaching the Social Studies.* 1927.

Judd, Charles H., *Proceedings of the Forty-Eighth Annual Meeting of the Associated Academic Principals of the State of New York.* 1930.

Kilpatrick, William H., *Education and the Social Crisis.* 1932.

Report of the Commission on the Social Studies, *Conclusions and Recommendations of the Commission.* 1934.

[2] Dunn, Arthur W., *The Social Studies in Secondary Education,* p. 9. U. S. Bureau of Education, Bulletin, 1916, No. 28.

[3] Rugg, E. U. and Dearborn, N. H., *Social Studies in Teachers Colleges and Normal Schools,* p. 15. 1928.

[4] Report of the Commission on the Social Studies, *op. cit.,* p. 6.

[5] Kimmel, William G., *Instruction in the Social Studies,* p. 1. U. S. Office of Education, Bulletin, 1932, No. 17.

[6] Beard, *op. cit.,* pp. 13-21.

[7] Dawson, and Others, *op. cit.,* pp. 56-87.

DEVELOPMENT OF THE PROBLEM

The problem here is concerned with the realization that pupils who undertake the study of the social sciences in secondary schools must have adequate preparation for the intelligent comprehension of these studies and adequate preparation for continuing the study of the social sciences in colleges and universities. This study undertakes to investigate to a limited extent the mathematics involved in the study of the social sciences and the provision that is made for learning that mathematics. It also endeavors to determine the connections between the mathematics and the social studies materials.

PREVIOUS CONSIDERATIONS OF THE PROBLEM

The National Committee on Mathematical Requirements, believing that it is essential that college entrance requirements must "insure adequate mathematical preparation" for the study of the social sciences in college, conducted an investigation that sought information to determine the content of the high school courses in mathematics that might serve as desirable preparation for the study of the social sciences in college.[8] The information was secured by asking a selected number of college teachers prominent in their particular fields to assign to each of a selected list of forty-one topics an estimate of its value as preparation for elementary courses in their respective fields.

The sixteen topics which were ranked highest by the social studies teachers are presented in Table I. The first column (N) contains the number of replies received for each topic. The remaining columns contain the percentages of the total number of replies for each topic that regarded the topic as either essential (E), of considerable value (C), of some value (S), or of little or no value (O) as preparation for the social studies.

A few of those replying considered graphs, the quadratic equation, variation, logarithms, computational tables, the slide rule, and the binomial theorem essential. Very little value was attached to the study of the topics of analytic geometry, numerical trigonometry, the theory of exponents, or problems leading to equations not shown in Table I.

[8] "The Reorganization of Mathematics in Secondary Education," *A Report of the National Committee on Mathematical Requirements,* pp. 44-45. 1923.

TABLE I

VALUE ATTACHED TO CERTAIN TOPICS IN MATHEMATICS AS PREPARATION FOR THE STUDY OF THE SOCIAL SCIENCES

Topic	N	E	C	S	O
Graphic representation of statistical data	21	57	24	14	5
Meaning and use of elementary concepts of statistics	22	55	36	5	5
Frequency distributions and frequency curves	21	47	33	10	10
Meaning and use of simple formulas	19	47	26	21	5
Meaning and use of negative numbers	18	45	17	22	17
Numerical computation with approximate data	12	40	27	20	13
Problems leading to linear equations in one unknown	15	40	7	20	33
Ratio and proportion	19	37	26	32	5
Correlation in statistics	21	33	47	14	5
Problems leading to simultaneous linear equations in two unknowns	12	33	8		58
Problems leading to quadratic equations in one unknown	13	31	8	8	54
Short-cut methods in numerical computation	17	29	35	23	12
The linear equation, $y = mx + b$	14	29	29	14	29
Arithmetic progression	17	23	29	12	35
Geometric progression	17	23	25	18	35
Probability	20	20	35	35	10

N = number; E = essential; C = considerable value; S = some value; O = little or no value.

In a somewhat similar study reported by Koch and Schlauch[9] five hundred college teachers were asked to "indicate in numeric order" the value of certain topics "by assigning number (1) to the most important topic, (2) to the topic second in importance, and so on through the list," and to place an *e* against those topics which they considered essential and an *x* against those topics which they would like to see omitted from secondary education. Since the study was undertaken by the Mathematics Section of the New York Society for the Experimental Study of Education in order to verify the findings of the National Committee, the topics included were similar to those used in the investigation of the National Committee, with additional topics on measurement of areas, measurement of volumes, checking of results, similarity of figures, mathematics of annuities, and permutations and combinations. The report, based on seventy-seven sufficiently complete replies, showed the estimates of the composite of all teachers, including

[9] Koch, E. H. and Schlauch, W. S., "Topics in High School Mathematics." New York Society for the Experimental Study of Education, *Contributions to Education*, Vol. I, pp. 351-359. 1924.

teachers of economics, to have a correlation of 0.6 with the findings of the National Committee. Five of the seven topics added in this study were placed among the essential topics.

Thorndike and his associates investigated the problem further by measuring the number of linear inches using algebra and the number of linear inches examined in nine textbooks in the social sciences, including books in American history, European history, civics, and economics. Of the 24,500 linear inches examined, 410 were found to use algebra 109 separate times. The report showed that the entire use of algebra was confined to the statistical graph.[10]

In another study it was found that the mathematics used in a random selection of textbooks in history, civics, economics, and commercial geography published between 1901 and 1924 could be included under the topics of numerals, indefinite expressions, common fractions, decimal fractions, comparisons, dates, United States money, foreign money, denominate numbers, mathematical terms, problems, diagrams and graphs, algebraic symbols, signed numbers, formulas, algebraic problems, and geometrical terms.[11]

More recently the Social Science Research Council has secured the report of a committee appointed to concern itself with the mathematical training desirable for students of agricultural economics and rural sociology.[12] The committee reported that it was of the opinion "that the needs of students of agricultural economics and rural sociology for training in collegiate mathematics are not essentially different from the needs of students who specialize in other branches of economics and sociology, and that all students would find some work in collegiate mathematics very helpful."

On the basis of this opinion the committee undertook to consider the needs of all students of the social sciences. It concluded that "for some social science students there is a fair question whether substantial direct benefit would be realized from any college mathematics whatever. On the other hand, some few students might profitably take even more mathematics than is involved in an ordinary undergraduate major in mathematics." The Committee further stated a belief that "a knowledge of the phases

[10] Thorndike, E. L. and Others, *The Psychology of Algebra*, p. 71. 1923.

[11] Cragg, M. E., *The Mathematics Found in the Social Science Textbooks*. 1925.

[12] "Collegiate Mathematics Needed in the Social Sciences." A Report in the *American Mathematical Monthly*, XXXIX, pp. 569-577, December, 1932.

of mathematics listed below would be helpful and useful to a large proportion of students of economics." "The training in such mathematics would help the student to acquire facility in reducing to definite terms the concepts and relationships encountered in the courses in the social sciences and to do clear and logical thinking and reasoning."

Topics in Mathematics Useful in Economics

1. Logarithms	Common logarithms; numerical computation; compound interest and annuities; Napierian logarithms.
2. Graphs	Plotting on ordinary and logarithmic paper; measurement of slopes and areas; graphical determination of maxima and minima; mean values; representation of three-variable relationships by means of three-dimensional diagrams and contours.
3. Interpolation	By reading ordinary or logarithmic graphs; by proportional parts; by successive differences.
4. Equations and forms of curves	Basic mathematical functions, their graphs, and characteristic properties of the variations represented; straight lines; parabolic and hyperbolic curves; exponential and logarithmic curves; logistic curves; sine and cosine curves.
5. Probability	Combinations; binomial theorem, elementary probability, the normal probability curve—its form, table, and equation; probability and frequency distributions; probability and time series.
6. Elements of differential and integral calculus	The significance of a derivative as a limit, as a rate or slope, as a frequency, etc.; the differential as an approximate increment; formula for differentiating elementary mathematical functions, u^n, a^u, log u, cos u, uv, u/v; partial differentiation; procedure for determining maxi-

	mum and minimum values; integration as the reverse of differentiation; relation between integration and summation; multiple integration.
7. Curve fitting	Plotting tabulated values, their logarithms, reciprocals, etc.; to ascertain whether the tabulated points lie on a curve of simple form; determination of the curve through such points, or selected points within the scatter diagram or chart; the method of least squares or other methods for obtaining the constants of curves of best fit.

CONCLUSIONS FROM PREVIOUS CONSIDERATIONS OF THE PROBLEM

The investigations concerned with the relationships that exist between mathematics and the social studies have been characterized by differences in purpose and by consequent differences in method of procedure.

"The nature of mathematics is twofold. On the one side it deals with quantitative relationships between material objects and thus becomes a tool in the world of business, economics, or science. On the other side it deduces theorems from arbitrarily chosen postulates and seeks to carry these theorems to their logical conclusions, giving little heed to their use in the world of practical affairs. From this point of view mathematics belongs to philosophy." [13] The social sciences treat of the human relationships of human beings, including within their province the entire range of human history. Desirable training in mathematics as preparation for the study of the social sciences may be limited to the study of concrete situations in mathematics for the purpose of acquiring the ability to recognize and intelligently understand similar situations when they are encountered with similar mathematical concepts in the literature of the social studies. On the other hand, desirable training in mathematics may be interpreted in a more general way to mean the acquisition of the ability to understand relationships in mathematics and the method of mathematics in dealing with relationships in order to apply the method in dealing with social re-

[13] Sanford, Vera, *A Short History of Mathematics,* p. 1. 1930.

lationships where mathematical concepts do not appear but where the mathematical method is equally efficacious. Mathematics is well suited to give this type of training. "Mathematics furnishes the only medium which the human mind has yet discovered for drawing conclusions that necessarily follow from set conditions, and algebra is the science which typifies and parallels the relationships of changing values which every person must recognize and deal with every day if he is to be a contributing and not a parasitical member of society. . . . Algebra is primarily a mode of thinking, offering to every inquiring mind an instrument of precision in interpreting and understanding world phenomena."[14] Much of geometry is equally valuable in exhibiting a logical structure of thinking and in affording a systematized means for developing the power of logical thinking. The teacher of mathematics "is more concerned with inculcating the virtues of straight thinking than with giving the individual the advantages that accrue from information."[15] It will require much experimentation rather than opinion to determine the extent to which mathematical training of this nature lends profit to the study of the social sciences, and to decide upon the topics in mathematics that are best adapted to provide such training.

In the discussion of the possibilities offered by mathematics for training in correct habits of thought it is not intended to underestimate the need for an intelligent understanding of the meanings involved in the mathematical concepts, facts, and processes that increasingly are finding their way into the social science materials. But this mathematics must be viewed, not as compartmentalized subject matter, but as a terse means of clarifying the exposition of human living and as a part of the vast body of integrated knowledge. Integration is a familiar term to both the mathematician and the economist. To the one it means a process of summation; to the other, the tendency to bring together under one management all the processes in the production of an article. In education its meaning is analogous. Integration connotes all that is held "essential in terms of adaptability, resourcefulness, knowledge, whole-

[14] Everett, John P., "Algebra and Mental Perspective." *Seventh Yearbook of the National Council of Teachers of Mathematics,* p. 26. 1932.

[15] *Ibid.,* p. 19.

some character, favorable attitudes, or skill. In short, it connotes all that goes to make up the general notion of oneness with the world or at-homeness in the world. It is correctly used to describe the behavior of persons after a certain kind of instruction has been carried on."[16] There is common agreement that departmentalized subject matter conceived as such is unlikely to promote this sort of behavior. From an educational standpoint the mathematical concepts, facts, and processes that occur in the literature of the social sciences, constitute just one phase of the relationship between mathematics and the social studies. Of no less importance are the connections in which the mathematics occurs in the social studies and the place of the correlated mathematics and social studies in the continuity or sequence of subject matter organized to attain the fullest measure of integration. The correlation of mathematics and the social studies can be studied by examining the content of the social studies. The development of much of the broader program will depend upon vision, judgment, and sound philosophy.

The changing nature of the social science materials and the wide range of the materials included must be considered in interpreting the findings of previous investigations. The materials of the social studies continually require reorganization and new modes of explanation to interpret changes in the social order. Many social situations involving mathematics do not get into the social studies materials. Some few are found in the content of social arithmetics; others appear in no printed materials. Graphs are finding a very definite place in the newer social studies materials. Further usage of mathematics seems certain. "Just as the multiplication table is applicable in more than one field of knowledge, so mathematics in general is peculiar to none. Sooner or later, every true science tends to become mathematical. The social sciences are simply a little later to be reached than astronomy, physics, and chemistry. . . . The outlook is bright for a healthy development of mathematics in the social sciences."[17] Finally, there is a correlation not popularized, because "though the simpler applications of mathematics made by competent writers and lecturers can be

[16] Featherstone, William B., "An Integrated Curriculum in the Secondary Schools." *California Quarterly of Secondary Education*, VIII, p. 340, June, 1933.

[17] Fisher, Irving, "The Application of Mathematics to the Social Sciences." *Sixth Yearbook of the National Council of Teachers of Mathematics*, p. 17. 1931.

appreciated by any intelligent readers and students, the more complicated analyses are only within the power of those who have mathematical ability."[18]

In previous investigations some attention has been given to the mathematics considered as desirable training for the study of the social sciences and to some extent to the correlation. Very little attention has been given to the mathematical needs of the social studies at different grade levels or of the various divisions of the social studies. Moreover, the connections in which mathematics appears have not been presented systematically.

PROBLEM AND PROCEDURE DEFINED

The present study undertakes to determine:

1. The mathematics and the mathematical connections that occur in the materials of instruction in certain prescribed courses in the social studies for the different grades at the junior high school level.

2. The provision that is made in mathematics textbooks for learning the mathematics that occurs in the junior high school social studies materials.

3. The mathematics and the mathematical connections that occur in the materials of instruction for certain introductory courses in the social studies at the college level.

4. The relation between the mathematics found in the prescribed courses and that occurring in the materials for elective courses in economics.

The method of selecting and analyzing the materials of instruction for the junior high school social studies and the results of the analysis of the materials are presented in Chapter II. Chapter III contains a report of the findings of an analysis to determine the extent to which textbooks in arithmetic and junior high school mathematics prepare for the intelligent comprehension of the mathematics in the social studies. Chapters IV and V are devoted to a report of the mathematics found in the materials for prescribed introductory courses in the social studies at the college level and for elective courses in elementary economics for high schools and colleges. Chapter VI gives a summary of the findings.

[18] Bowley, A. L., *The Mathematical Groundwork of Economics*, p. v. 1924.

CHAPTER II

MATHEMATICS IN SELECTED JUNIOR HIGH SCHOOL SOCIAL STUDIES

SELECTION OF THE SOCIAL STUDIES MATERIALS

THE diversity of meanings that have been attached to the social studies, the varied plans of organization and combination of subjects, and the numerous textbooks and other reference materials that are available necessitated limiting the materials of instruction to be examined at the junior high school level to certain materials representing selected types of organization in prescribed courses. The selection of these materials was based on the findings of certain previous studies of the content of courses and of the materials in use.

From a study of the social studies offered at the junior high school level in fifty-five cities, it was concluded that "the courses of study examined, to the extent that they may be regarded as typical, seem to indicate that fusion or unified courses are received with increased favor at the junior high school level. The content in these courses is predominantly geography, history, and civics for grades 7, 8, and 9, respectively, regardless of attached labels." [1]

Another survey of the social studies courses in 301 junior high schools showed geography, American history, and the Rugg fusion course as most frequently prescribed for the seventh year; American history, civics, geography, and the Rugg fusion course for the eighth year; and civics, ancient and world history, and the Rugg fusion course for the ninth year.[2]

Coxe concluded from his study in New York State that geography is most likely to be taught in the seventh grade, history is

[1] Kimmel, William G., *Instruction in the Social Studies*, pp. 13-16. U. S. Office of Education, Bulletin, 1932, No. 17.

[2] Wilson, Howard E. and Erb, B. P., "A Survey of Social Studies Courses in 301 Junior High Schools." *School Review*, XXXIX, pp. 497-507, September, 1931.

confined to the seventh and eighth grades, and civics is usually a ninth grade subject.[3]

Wilson summarizes the status of the social studies at the junior high school level as follows: "As had been the case for some years before 1925 and still is the case today, geography was offered more frequently than any other course in Grade VII; United States history led in Grade VIII; and civics, though not decisively, was offered more frequently than any other course in Grade IX."[4] Wilson asserts further with reference to the previous survey of the social studies courses in 301 junior high schools that "approximately a fourth of the schools reporting indicated that they were experimenting, more or less whole-heartedly, with fusion courses, and the interest in such courses was even more widespread,"[5] and that "the Rugg course is by all odds the most widely used fusion course in the United States."[6]

In order that the results of this part of the study may represent the actual mathematics encountered by groups of junior high school pupils in the materials adapted for their prescribed courses in the social studies, a syllabus was selected that appeared to be in substantial agreement with the conclusions in the studies to which reference has been made.[7] The syllabus contained a complete sequence for the three-year period and provided references for each topic. To adapt the instruction to the varying abilities of the pupils, a flexible two-track plan was used, and separate references were given for pupils of superior ability and for pupils of limited ability. Pupils of average ability were included in one or the other of these two groups. For the purposes of this study the syllabus provided the additional advantage that, in general, courses in separate subjects are offered for slow pupils, while the average and most capable pupils are using a fusion course. That reference was chosen for each group of pupils that was suggested in the syllabus for the largest number of topics. In general,

[3] Coxe, Warren W., "When Is a School a Junior High School?" *Junior-Senior High School Clearing House,* V, pp. 72-79, October, 1930.

[4] Wilson, Howard E., *The Fusion of the Social Studies in Junior High Schools,* p. 3. 1933.

[5] *Ibid.,* p. 4.

[6] *Ibid.,* p. 39.

[7] *Social Science, Junior High School, Grades Seven, Eight, and Nine.* Public Schools, Denver. Course of Study Monograph, No. 3. 1931.

this procedure resulted in the selection of one textbook, or a portion of one, for each group of pupils for each half year. For purposes of comparison references from another source were selected.[8] The complete list of books examined, with the number of pages examined in each, is given in Appendix I.

Although by no means all the types of junior high school social studies courses and materials will be considered, the findings from the analysis of the materials selected and appearing on the pages that follow will undoubtedly justify the conclusion that the examination of other social studies materials for the prescribed courses at the junior high school level would show the type of mathematics occurring in the materials to vary to a lesser degree than the type, organization, or range of social studies materials.

THE TERM MATHEMATICS AS APPLIED TO THIS STUDY

The procedure followed in determining the part of the content of the social studies materials to be classed as mathematics was based on the twofold nature of mathematics described in Chapter I. This study is limited to that phase of mathematics that deals with quantitative relationships; the existence of the philosophical phase will not be considered.

"Every quantitative relationship involves one of two questions, either how many or how much. The answer to the former leads to counting and the latter to measuring."[9] A proper understanding of measurement implies a consideration of the approximate nature of all measurement and of the units employed in measuring. The term mathematics, as used in this study, is meant to include those words, figures, and other symbols that have been adapted to describe the results of counting and measuring and the methods and language that have been developed in the course of manipulating, comparing, and presenting the results and processes involved in counting and measuring.

RESULTS OF THE ANALYSIS OF THE SOCIAL STUDIES MATERIALS

The first seven tables in this chapter are designed to show the use of number and the connections with which number is used.

[8] Fancler, D. G. and Crawford, C. C., *Teaching the Social Studies.* 1932.
[9] Schaaf, W. L., *Mathematics for Junior High School Teachers,* p. 1. 1931.

Each separate use of mathematics and its corresponding connection with the social studies were tabulated. The frequencies have been totaled by grades for the Rugg fusion materials and for the other materials for each concept classed as mathematical. In comparing the mathematics in the fusion materials with that in the others, the relative frequencies in the tables cannot be considered significant, since the number of pages examined in each case differs extensively. The tables can be used to compare the type of mathematics found in each case and to compare the relative use of two or more concepts by means of the total frequencies.

The frequency of occurrence of ordinal numbers is shown in Table II. Ordinal numbers occur expressed in Arabic figures, in words, and in Roman notation. There is some tendency throughout to express such numbers in figures. Those expressed in words are predominantly "first" and "second," with an occasional use of a larger number, such as might be used in designating the "Seventy-second Congress." The Roman numerals that occur are formed primarily from the symbols I, V, and X. But Roman numerals containing other symbols are so well distributed that the principles involved in their formation should be understood.

Dates in years occur for A.D. and for B.C., and are given as single years in such forms as 1930 or '76; and for intervals in the form 1841–1846. Periods are expressed by such phrases as the 1900's, and 1900s, the 90's, the 90s, the nineties. Various ways are used to state time more exactly. These include such expressions as April 6, 1840; 26th of June, 1930; twenty-third of April, 1789; July 1-3, 1863; seven o'clock; 2:30 o'clock; 2.30 o'clock; 7 A. M.; 5 a. m.; 3 P. M.; and 4 p. m.

The connections for all the ordinal numbers in Table II that were expressed in figures and in Roman notation are shown by grades in Table III. An ordinal number denotes a certain one in a series of objects or things. By ordinal number connection is meant the name of the object or thing. In a similar way connections with other mathematical concepts will be listed.

The 11,822 different occurrences of ordinal numbers in figures were used in seventeen different ways, only five of which were common in all three grades. The five included page numbers, references, statements of facts, hours and minutes on the clock, and

TABLE II

FREQUENCY OF OCCURRENCE OF ORDINAL NUMBERS IN THE JUNIOR HIGH SCHOOL SOCIAL STUDIES MATERIALS EXAMINED

Ordinal Number	Fusion Materials		Other Materials		Total
	Grades		Grades		
	7	8	7	9	
In Figures					
One figure	333	1,619	1,729	540	4,221
Two figures	406	992	898	324	2,620
Three figures	951	2,124	1,127	766	4,968
Four figures	6	3		2	11
Five figures		1		1	2
Subtotal	1,696	4,739	3,754	1,633	11,822
In Words					
Units	251	877	424	154	1,706
Tens	26	56	59	19	160
Subtotal	277	933	483	173	1,866
In Roman Numerals					
I–VIII	96	341	88	185	710
IX–XXXIX	83	152	52	64	351
Over XXXIX	6	5	2	1	14
Subtotal	185	498	142	250	1,075
Dates					
Year only	1,372	3,977	293	215	5,857
Period	105	283	2		390
Year and month	168	135	2	10	315
Year, month, and day	55	147	4	16	222
Month and day	4	65	2	8	79
Subtotal	1,704	4,607	303	249	6,863
Total	3,862	10,777	4,682	2,305	21,626

topics in outline. Roman numerals were used in fourteen different ways. The uses common to all grades were to number chapters of a book, to number parts of an exposition, and to number topics in outline. The largest, or five-place numbers, were used to indicate serial numbers on certificates of money.

The frequency, range, and nature of the cardinal numbers that occurred are shown in Table IV. The range was from 0 to 350,000,000,000 in figures, and from zero to ten billion in words. The frequencies include all cardinal integral numbers, the integral parts of mixed numbers, and single, double, triple, and similar terms. Two thirds of the 12,817 numbers were found representing numbers less than one hundred, and one and one third per cent

TABLE III

ORDINAL NUMBER CONNECTIONS IN THE JUNIOR HIGH SCHOOL SOCIAL STUDIES MATERIALS EXAMINED

Connection with Numbers in Figures	Grade 7	8	9	Connection with Numbers in Roman Notation	Grade 7	8	9
Chapter		x	x	Amendment			x
Day		x		Article		x	x
Figure	x	x		Chapter	x	x	x
Page	x	x	x	Direction			x
Paragraph			x	Group	x		
Parallel		x	x	Page	x	x	
Problem	x		x	Part	x	x	x
Project	x			Problem		x	
Question	x		x	Royalty	x	x	
Reference	x	x	x	Section		x	x
Section		x		Table	x		
Serial			x	Topic in outline	x	x	x
Statement	x	x	x	Unit	x	x	
Street		x	x	Volume	x	x	
Time	x	x	x				
Topic in outline	x	x	x				
Volume		x					

were in the billions. There is a tendency to express small cardinal numbers and indefinite numbers in words. Numbers written in the form "120 millions," "400 thousands," and in similar forms occur in tables and on graphs. The range and frequencies of cardinal numbers do not vary significantly in the different grade materials at the junior high school level. The differences are accounted for more by the characteristic style of the author than by the nature of the materials or the grade level.

The thirty-five different valued fractions that were expressed in sixty different ways are listed with their frequencies in Table V. Most of the fractions were expressed in words. Two thirds of the 723 uses were with the fractions three fourths, two thirds, one half, and one third. Ninety-seven per cent of the denominators were smaller than ten. Occasionally fractions not in their lowest terms were used when it apparently seemed desirable to retain like denominators to make comparisons more evident.

The 141 different connections with fractions, the 843 connections with cardinal numbers, and the 367 connections with ordinal

TABLE IV

FREQUENCY OF OCCURRENCE OF CARDINAL NUMBERS IN THE JUNIOR HIGH SCHOOL SOCIAL STUDIES MATERIALS EXAMINED

Cardinal Number	Fusion Materials Grades 7	Fusion Materials Grades 8	Other Materials Grades 7	Other Materials Grades 9	Total
In Figures					
One figure	120	135	279	140	674
Two figures	507	560	817	135	2,019
Three figures	311	359	501	35	1,106
Four figures	140	224	79	16	559
Five figures	57	190	22	23	292
Six figures	67	121	22	36	246
Seven figures	132	131	17	35	315
Eight figures	115	113	6	22	256
Nine figures	63	69	5	17	154
Ten figures	28	49		6	83
Eleven figures	2	23		3	28
Twelve figures	2	7			9
Subtotal	1,544	1,981	1,748	468	5,741
In Words					
Units	683	2,334	1,201	844	5,062
Tens	104	417	175	154	850
Hundreds	57	126	176	30	389
Thousands	51	167	103	20	341
Ten thousands	1	33	3	5	42
Hundred thousands	14	22	10	3	49
Millions	69	118	57	12	256
Ten millions	2	6	3		11
Hundred millions	9	6	8	1	24
Billions	16	29		1	46
Ten billions	4	2			6
Subtotal	1,010	3,260	1,736	1,070	7,076
Total	2,554	5,241	3,484	1,538	12,817

numbers in words are not included in the tables. The connections are similar in nature to those listed in Table XII.

Decimals occurred less frequently than common fractions. The range, with the exception of a single quotient obtained in the process of calculating a tax rate, was confined to tenths and hundredths. The tax rate was carried to thousandths. Three books contained no decimals, and three others contained two, four, and five respectively. Less than six per cent of the decimals were pure decimals. Averages, per cents, and the common method of writing

TABLE V

FREQUENCY OF OCCURRENCE OF FRACTIONS IN THE JUNIOR HIGH SCHOOL SOCIAL STUDIES MATERIALS EXAMINED

Fraction	Fusion Materials		Other Materials		Total
	Grades		Grades		
	7	8	7	9	
19/20				2	2
Nine tenths, 9/10		4	7	4	15
Seven eighths, 7/8		1		1	2
Five sixths	2				2
Four fifths, 8/10, 4/5	4	1	3	5	13
Three fourths, three quarters	8	13	10	8	39
7/10				1	1
Two thirds, 2/3	4	36	14	21	75
Three fifths, 6/10	2	3		3	8
7/12			2		2
One half, half, 5/10, 1/2	52	151	81	31	315
Two fifths, four tenths, 4/10		1	4	4	9
One third, third, 1/3	25	34	19	8	86
Three tenths, 3/10			4	1	5
One fourth, fourth, one quarter, quarter, 1/4	22	29	9	4	64
One fifth, two tenths	5	12	2	2	21
One sixth, 1/6	3	7	2	1	13
One seventh	2	1	1	1	5
2/15			1		1
One eighth, eighth		4			4
One ninth	1				1
One tenth, tenth, 1/10	5	11	1	1	18
One twelfth	1				1
One fourteenth	1				1
One fifteenth		3			3
Sixteenth	1				1
One twentieth, twentieth		3	5		8
One twenty-fifth	1				1
One twenty-seventh			1		1
One thirtieth		1			1
One thirty-second	1				1
One fiftieth			1		1
Hundredth			1		1
1/45	1				1
Three millionths		1			1
Total	141	316	168	98	723

TABLE VI

FREQUENCY OF OCCURRENCE OF DECIMALS IN THE JUNIOR HIGH SCHOOL SOCIAL STUDIES MATERIALS EXAMINED

Decimal	Fusion Materials		Other Materials		Total
	Grades		Grades		
	7	8	7	9	
Tenths					
Significant figures					
One		2		1	3
Two	5	29		3	37
Three	10	73		22	105
Four		18			18
Subtotal	15	122		26	163
Hundredths					
Significant figures					
One	2				2
Two	4			11	15
Three	46	16		25	87
Four	10	7		62	79
Five		3		2	5
Six				3	3
Subtotal	62	26		103	191
Thousandths					
Significant figures					
Two				1	1
Subtotal				1	1
Total	77	148		130	355

dollars and cents accounted for most of the uses. The nature and frequency of the decimals in the social studies materials are shown in Table VI and the twelve connections with decimals in Table VII.

A large proportion of the numbers in the social studies materials are approximate numbers. In seven cases the numbers were described as round numbers. Indefinite numbers, such as "hundreds of," "thousands of," and others are in frequent use. In other cases the qualifying words listed in Table VIII were used to indicate approximate numbers. One out of every ten cardinal numbers was accompanied by one of these words.

A few illustrations will show more exactly the use of cardinal numbers in the materials examined.

1. The total cost was less than $186,000,000,000.[10]

[10] Rugg, Harold, *Changing Civilizations in the Modern World*, p. 372. 1930

TABLE VII

CONNECTIONS WITH DECIMALS IN THE JUNIOR HIGH SCHOOL SOCIAL STUDIES MATERIALS EXAMINED

Connection	Grade 7	Grade 8	Grade 9
Average number of people per square mile		x	
Average number of people			x
Average number of workers		x	
Cents			x
Dollars	x	x	x
Per cents in tables (e.g., exports, imports, etc.)	x	x	
Per cents of people		x	
Per cents on bar graphs			x
Per cents on circle graphs	x	x	
Quotient (result of calculating a tax rate)			x
Ratio			x
Relative prices		x	

TABLE VIII

FREQUENCY OF WORDS USED TO INDICATE APPROXIMATE NUMBERS IN THE JUNIOR HIGH SCHOOL SOCIAL STUDIES MATERIALS EXAMINED

Word	Grade 7	Grade 8	Grade 9	Total
About	181	184	21	386
Above		3		3
Almost	21	7		28
Approximately	8	37		45
At least	9	13	16	38
Barely		2		2
Estimate	16	50	5	71
Less than	68	60	6	134
More than	177	167	18	362
Nearly	78	64	5	147
Over	86	48	18	152
Practically	1			1
Round number	2	5		7
Under	13	1		14
Total	660	641	89	1,390

2. The round number 1,900,000,000 as an approximate estimate of the total population of the earth.[11]

3. France isn't one twenty-fifth as large as the whole empire.[12]

4. Germany, 1/45 the size of Russia.[13]

The units of measure employed in the social studies materials, with the frequencies of their occurrence, are listed in Table IX. Sixty-five different units were employed. Only the year, the month, and the day were common to all the textbooks examined. The common units of time have the largest frequencies. The large frequencies for the degree of arc arise from its use on maps to indicate degrees of longitude and latitude. The five other most frequently used units are the foot and the mile as units of length, the acre and the square mile as units of area, and the ton as a unit of weight. In several instances the equivalence of some more uncommon units, including foreign monetary units and units of length, such as the league, were shown in parentheses or brackets in terms of more commonly used units.

The connections with which the units of measure were used were segregated according to the type of unit and, where possible, combined and listed as a class in Table X. Thus, under units of capacity, foods represent a large number of different items that occurred in the textbooks examined. The connections with units of counting, of time, and of temperature were either so diverse or so well known that they are not included in the table.

The more exact nature of the connections with units of measure can be observed in the following illustrations taken from the materials examined:

1. It (the observatory in the part of London known as Greenwich) is well known the world over because it marks the prime meridian of longitude, or the meridian of 0°. You will remember that the meridians are the imaginary north-south lines which pass around the earth through the poles. They are numbered by degrees east and west from Greenwich. Since there are 360 degrees in a circle, the meridian of 180° is just halfway around the earth in either direction, west or east, from Greenwich. Use a globe to prove that this is true.[14]

[11] *Ibid.*, p. 5.
[12] *Ibid.*, p. 196.
[13] *Ibid.*, p. 371.
[14] Atwood, Wallace W. and Thomas, Helen G., *Nations Beyond the Seas, Book Three*, p. 24, 1930.

TABLE IX

FREQUENCY OF OCCURRENCE OF UNITS OF MEASURE IN THE JUNIOR HIGH SCHOOL SOCIAL STUDIES MATERIALS EXAMINED

Unit of Measure	Fusion Materials		Other Materials		Total
	Grades		Grades		
	7	8	7	9	
Acre	30	80	22	3	135
Barrel	4	11	1	1	17
Board foot		3			3
Bushel	11	17	6	3	37
Cent	4	46		24	74
Century	52	79	165	11	307
Crown		2			2
Cubic foot				1	1
Cubit	1		1		2
Cwt.				2	2
Day	122	431	185	93	831
Decade	3	24		16	43
Degree (arc)	322	146	969		1,437
Degree (temperature)	2	3	14	1	20
Dime				1	1
Dollar	63	469	10	112	654
Double-eagle				1	1
Dozen	6	10	3	11	30
Dram			1		1
Eagle				1	1
Fathom	4	3	1		8
Five-cent				1	1
Foot	31	25	75	12	143
Franc				2	2
Gallon	2	14	4	2	22
Grain			2	1	3
Guilder		1			1
Guinea		1			1
Half-dollar				1	1
Half-eagle				1	1
Hogshead		3			3
Horsepower	2	1		1	4
Hour	38	96	20	48	202
Inch	1	17	53	5	76
Kilowatt-hour		1			1
Knot		2			2
League	1	16	2		19
Long-ton	3			1	4
Mark				1	1
Metric-ton				2	2
Mile	126	190	72	23	411

TABLE IX (Continued)

Unit of Measure	Fusion Materials		Other Materials		Total
	Grades		Grades		
	7	8	7	9	
Minute (arc)		9			9
Minute (time)	2	20	3	10	35
Month	129	241	38	20	428
Nautical mile		1			1
Ounce	2	2	2		6
Peck		1			1
Pence		3			3
Penny		4		1	5
Peso				1	1
Pound (sterling)	2	12	2	2	18
Pound (weight)	17	36	1	6	60
Quart		1		8	9
Quarter				2	2
Score	1	1		3	5
Second (time)			2		2
Shilling	2	9		2	13
Short-ton		4		2	6
Square foot		1	10		11
Square mile	40	82	44	6	172
Ton	54	61	4	5	124
Week	11	49	26	11	97
Yard	2	16		5	23
Year	343	1,025	403	208	1,979
Yen	1			1	2
Total	1,434	3,169	2,141	675	7,419

2. The latitude 36° 30′ marked the limit north.[15]

3. Some of the railroads have a narrow gauge of 3 feet, 6 inches; others a "standard gauge" of 4 feet, 8½ inches, as we do in the United States; and still others a broad gauge of 5 feet, 3 inches.[16]

4. The water is 100 fathoms, or 600 feet, deep.[17]

5. One might journey eight leagues (about 25 miles).[18]

6. Soon after the law was passed, government surveyors went into the Ohio Valley and measured and staked out land. First, townships were laid out, each one six miles square, totaling 36 square miles of land. Secondly, each township was subdivided into 36 sections of one square mile each, that is, 640 acres. Third, each section was divided into four quarter-sections of 160 acres each.[19]

[15] Rugg, Harold, *Changing Civilizations in the Modern World*, p. 125. 1930.
[16] Atwood, Wallace W. and Thomas, Helen, *op. cit.*, p. 114.
[17] *Ibid.*, p. 18.
[18] Rugg, Harold, *op. cit.*, p. 45.
[19] Rugg, Harold, *A History of American Civilization*, p. 307. 1930.

TABLE X

CONNECTIONS WITH UNITS OF MEASURE IN THE JUNIOR HIGH SCHOOL SOCIAL STUDIES MATERIALS EXAMINED

Type of Unit and Connection	Grade 7	8	9	Type of Unit and Connection	Grade 7	8	9
Units of arc				Units of money			
Equator	x	x		Admission		x	x
Latitude	x	x		Amount	x	x	x
Longitude	x	x		Appropriation			x
Meridian	x	x		Bond		x	
Units of capacity				Capital	x	x	
Foods	x	x	x	Cash	x		
Grain	x	x		Certificate			x
Liquids		x	x	Claim		x	
Medicine			x	Cost	x	x	x
Oil	x	x		Debt	x	x	x
Seed	x			Dividend		x	
Tobacco		x		Duty		x	x
Units of length				Earnings		x	x
Breadth		x	x	Expenses	x	x	x
Circumference	x			Exports		x	
Depth	x	x		Fare	x	x	
Diameter		x		Fee		x	
Distance	x	x	x	Fine		x	
Elevation	x		x	Fortune		x	
Gauge	x			Gift	x		
Height	x	x		Goods	x	x	
Length	x	x	x	Gold	x	x	
Radius		x		Imports	x		
Rainfall	x	x		Income		x	x
Thickness	x	x	x	Investment		x	
Width	x	x	x	Loan		x	x
Units of surface				Loss	x		
Area	x	x	x	Money	x	x	x
Canvass	x			Note		x	x
Land	x	x	x	Price		x	x
Oval	x			Prize	x		
Units of weight				Property	x	x	
Armor	x			Rent	x	x	
Exports		x		Reparations	x		
Foodstuffs	x	x	x	Salary		x	x
Freight	x	x	x	Stock		x	
Fur		x		Sum	x	x	x
Goods	x	x	x	Tariff		x	
Grain	x			Tax	x	x	x
Metals	x	x	x	Trade	x	x	
Mail	x			Value	x	x	x
Minerals	x	x	x	Wages	x		x
Persons			x	Worth	x	x	
Pressure			x	Wealth			x
Ships		x	x				

7. It is not uncommon for the temperature to drop to 50°F. or 60°F. below zero there.[20]

8. At this time an ounce of gold was worth 15 times as much as an ounce of silver; so the ratio of 15 to 1 was used when the two metals were coined.[21]

9. It hauled a load of thirteen tons in wagons at an average speed of 35 miles an hour.[22]

Table XI contains ninety-seven terms used in a mathematical sense in the social studies materials according to the definition of mathematics previously stated. The list does not include the units of measure in Table IX, the mathematical terms associated with measurement in Table X, nor a large number of terms such as wider, more than, maximum, number of, majority, and many others that are used to make indefinite comparisons or that have been in such frequent use in earlier materials that their meaning is assumed to be understood. In order to have a definite criterion for excluding these terms, the eight series of arithmetic textbooks prepared for grades three through six inclusive and listed in Appendix II were examined to determine those terms that were used without definition. All such terms were excluded in preparing Table XI. Half the terms listed in the table are names given to arithmetic processes or terms associated with such processes. Most of the remaining terms are those associated with geometric figures or with statistical graphs.

Abbreviations for the commonly used units of measure are in general use. Symbols frequently occur for dollar, cent, pound sterling, per cent, degree, minute of arc, and Fahrenheit; to indicate addition, subtraction, multiplication, and division; and include the decimal point, colon, and dash.

The connections with which each of the terms in Table XI was used are not included in the tables. Tables XII, XIII, and XIV show the connections with per cent, average, and rate. The terms appear as in the following two illustrations:

1. Since there were on the average about one and one half workers in each family, the income per family in these representative occupations was approximately $2,100 ($1,426 x 1.5). This checks fairly well with

[20] Atwood, Wallace W. and Thomas, Helen, *op. cit.*, p. 272. 1930.
[21] Rugg, Harold, *A History of American Government and Culture,* p. 404. 1931.
[22] Rugg, Harold, *Changing Civilizations in the Modern World,* p. 75. 1930.

TABLE XI

MATHEMATICAL VOCABULARY IN THE JUNIOR HIGH SCHOOL SOCIAL STUDIES MATERIALS EXAMINED

Term	Fusion Materials		Other Materials		Total
	Grades		Grades		
	7	8	7	9	
Add(ed)	19	33	13	15	80
Algebra			1		1
Angular		1			1
Approximate(ly)	8	37		2	47
Arabic figures			1		1
Arc	2				2
Area	60	20	25	10	115
Arithmetic			2		2
Arithmetical	1	1		2	4
Average	10	51	5	8	74
Balance				1	1
Bar graph	6	1			7
Base		1			1
Calculate, calculation		2			2
Center	1		1		2
Check	1	1			2
Cipher			1		1
Circle	9	3	11	1	24
Circle graph	5				5
Circular		1	1	1	3
Circumference			2	1	3
Cylinder	1	1	2		4
Decimal			1	1	2
Diagonal				1	1
Diagonal line	1				1
Diameter		2			2
Difference			7		7
Discount				1	1
Divide(d), divisible by	12	57	17	15	101
Dividend	1	2		6	9
Edge			1		1
End	1	2			3
Equal(ly)	4	11	7	4	26
Equivalent		1			1
Estimate	15	58	1	9	83
Figure		2	1	1	4
Figures	10	11	1		22
Fraction	1	1		1	3
Gauge			4		4
Geometry			1		1
Graph	21	14		5	40
Hemisphere	1	3	24		28
Imaginary line	1		1		2
Infinity		1			1
Interest		12		12	24
Intersection				2	2

TABLE XI (Continued)

Term	Fusion Materials		Other Materials		Total
	Grades		Grades		
	7	8	7	9	
Line	14	26	30	11	81
Mathematical		1			1
Mathematician			1		1
Mathematics			2	3	5
Measure	2	6	7	4	19
Metric system				2	2
Multiplication		2			2
Multiply, multiplied	1	3	1		5
Net profit		1			1
Oblong	1				1
Oval	1				1
Parallel		3	4	3	10
Per cent	64	63	8	75	210
Percentage	8	6		9	23
Perpendicular		2			2
Pictograph	1				1
Plus		3		2	5
Point	2	3			5
Proportion(al)	12	23	6	9	50
Pyramid	3		11		14
Radius	1	1			2
Rate	3	28		33	64
Ratio		3		2	5
Reckon		1	1		2
Rectangular	1	1			2
Right angle		1			1
Round number	2	5			7
Roman numerals			1		1
Scale	6	8	46		60
Sector			1		1
Semicircle			1		1
Side			3		3
Solve	1				1
Spiral			2		2
Square	4	18	9	3	34
Statistics	2	8		1	11
Straight line	1	1	1		3
Sum	7	32	7	10	56
Table	26	9	1	6	42
Time line	2	1	6		9
Time schedule	4	1			5
Times	32	24	13	3	72
Triangle		1			1
Triangular		7			7
Unit	2	1		6	9
Varies		1	1		2
Volume	1	2			3
Zero				1	1
Total	395	626	294	282	1,597

the estimate of the average income per family for the poorest 90 per cent of the people.[23]

2. Fix a radius of 3,600 miles, and draw an arc like the one on the map, cutting across Asia from Japan to Western India. Now draw imaginary lines connecting the two ends of the arc with the center of Java.[24]

TABLE XII

CONNECTIONS WITH PER CENT IN THE JUNIOR HIGH SCHOOL SOCIAL STUDIES MATERIALS EXAMINED

Connection	Grade 7	Grade 8	Grade 9	Connection	Grade 7	Grade 8	Grade 9
Accidents			x	Jobs		x	
Acres	x			Labor			x
Area	x			Land	x	x	
Beef		x		Manufacturing		x	
Bonds		x		Meat	x	x	
Business	x		x	Money		x	
Capital		x		Oats	x	x	
Cattle	x			Oil		x	
Census	x			Output	x		
Children		x		People	x	x	x
Cities		x		Platinum	x		
Coal	x			Population	x	x	x
Collars		x		Potatoes	x		
Consumption	x			Production	x		
Corn	x	x		Products	x		
Crops	x			Profit		x	
Decrease	x	x		Property		x	x
Deposits			x	Pupils		x	
Dividend	x		x	Refineries		x	
Duties		x		Removals		x	
Earnings	x	x		Rye	x		
Expenses			x	Ships	x	x	
Exports	x			Silk	x		
Families		x		Skin		x	
Farms		x		Stenographers			x
Features			x	Sugar	x	x	
Food	x			Tax		x	x
Graduates			x	Trade		x	
Highways			x	Typists			x
Houses		x		Value		x	x
Immigrants		x	x	Vessels	x		
Imports	x			Voters		x	
Income		x		Wealth		x	
Increase	x	x	x	Wheat		x	
Inhabitants	x			Wool		x	
Interest		x	x	Workers	x	x	x
Iron	x						

[23] Rugg, Harold, *A History of American Civilization*, p. 606. 1930.

[24] Rugg, Harold, *Changing Civilizations in the Modern World*, p. 5. 1930.

TABLE XIII

Connections with Average in the Junior High School Social Studies Materials Examined

Connection	Grade 7	Grade 8	Grade 9	Connection	Grade 7	Grade 8	Grade 9
Cost in dollars		x		Salary in dollars		x	x
Day in hours		x		Speed in miles per hour	x		
Family in persons		x		Tax in dollars		x	
Farm in acres	x	x		Temperature in degrees	x	x	
Elevation in feet	x		x	Time in hours		x	
Income in dollars		x		Trade in dollars	x	x	
Length in inches		x	x	Unemployed in persons		x	
Life in years		x		Value in dollars	x	x	
Number of people	x			Wage in dollars		x	
Production in cubic feet			x	Weight in tons	x		x
Rainfall in inches		x		Work in hours		x	

TABLE XIV

Connections with Rate in the Junior High School Social Studies Materials Examined

Connection	Grade 7	Grade 8	Grade 9	Connection	Grade 7	Grade 8	Grade 9
Dividend		x	x	Cents a yard			x
Freight		x	x	Dollars an acre	x	x	
Increase		x		Dollars a day		x	x
Interest		x	x	Dollars an hour		x	
Profit		x		Dollars a month			x
Railroad		x		Dollars a week			x
Salary			x	Hours a day	x	x	
Speed			x	Hours per week			x
Tariff		x	x	Inches a year	x		
Taxes		x	x	Miles a day	x		
Acres a day	x			Miles an hour	x	x	x
Bushels per hour			x	Miles a minute		x	
Bushels a year	x			Persons per square mile	x	x	
Cents a barrel		x		Tons a year	x		
Cents a gallon		x					

The quantitative aspect of the social studies is evidenced further in the extensive use of tables and graphs. Eighty-five tables appeared in the junior high school social studies materials examined. Thirty-five of these appeared in the materials for the seventh year. The tables were used to present in detailed form the growth of populations, the growth in exports, the increase in the production of various products, and various other comparisons and increases. The majority of the tables were limited to two or three columns and less than seven items.

The graphs that appeared included horizontal and vertical bar graphs, curved and broken line graphs, circle graphs, and pictographs. Only three of the books examined contained no graphs. Table XV shows the distribution of the graphs in the materials examined. Table XVI shows the results of an analysis of all the graphs that occurred.

TABLE XV

FREQUENCY OF OCCURRENCE OF GRAPHS IN THE JUNIOR HIGH SCHOOL SOCIAL STUDIES MATERIALS EXAMINED

Type of Graph	Fusion Materials		Other Materials		Total
	Grades		Grades		
	7	8	7	9	
Bar graph	40	21		8	69
Circle graph	21	5		19	45
Line graph	2	5		3	10
Pictograph	4			2	6
Total	67	31		32	130

Figures 1-8 on page 31 show the nature of the backgrounds or rulings used for the graphs. A plain background with no ruling, as in Figure 1, was used occasionally with horizontal and vertical bar graphs. The background in Figure 2 was in most frequent use with vertical bar graphs and that in Figure 5 with horizontal bar graphs. Figures 3, 4, and 6 are similar to Figures 2 and 5, with some subdivisions indicated. Figures 7 and 8 represent rulings commonly used with broken and curved line graphs.

The variations in applying scales to the graphs are shown in Figures 9-20. Discrete quantities other than those shown in Figures 12, 19, and 20 were in frequent use.

The more frequent of the graph types are emphasized in Figures 21-30, where the rulings and the scales have been omitted from the figures. The types shown in Figures 21, 24, and 30 occurred most frequently. The bar graphs often occurred with the bars grouped, as shown in Figures 23, 25, and 26, and occasionally divided, as in Figure 22. Figures 27 and 29 are typical of the type of broken

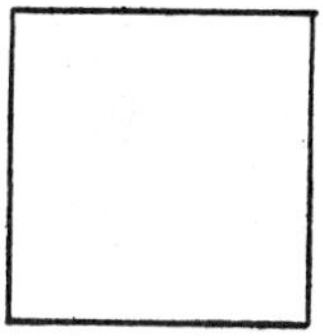

Figure 1

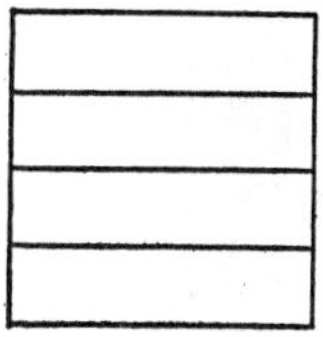

Figure 2

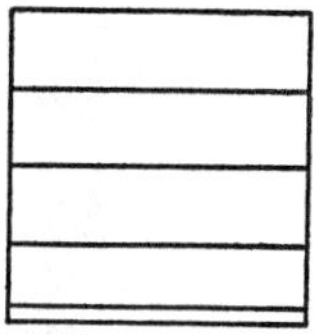

Figure 3

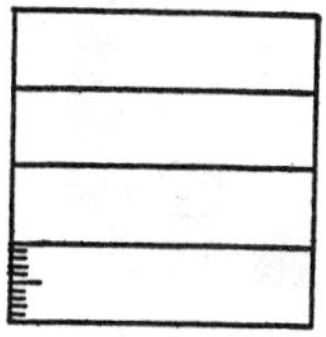

Figure 4

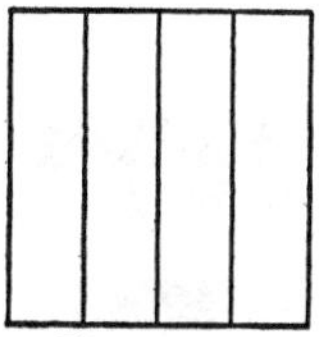

Figure 5

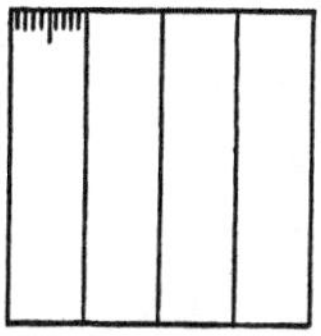

Figure 6

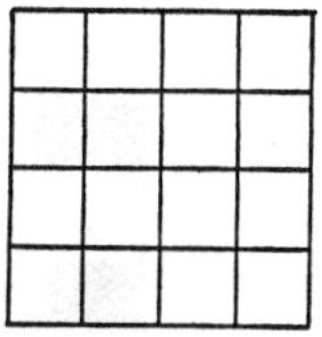

Figure 7

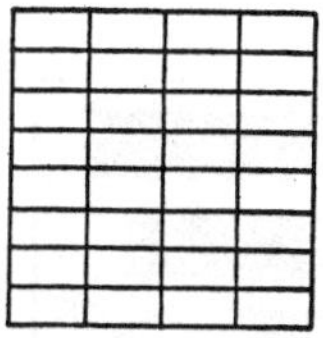

Figure 8

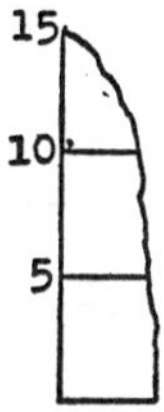

Figure 9

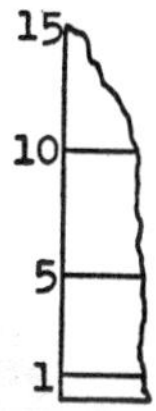

Figure 10

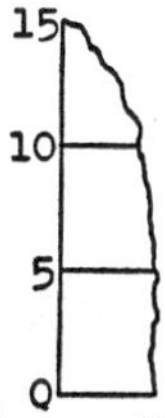

Figure 11

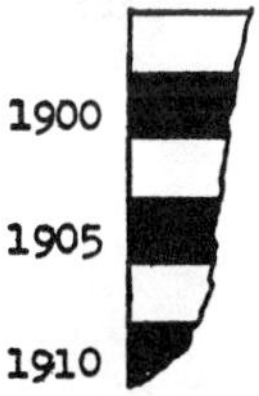

Figure 12

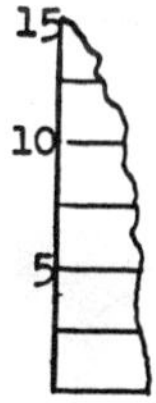

Figure 13

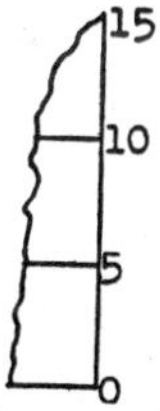

Figure 14

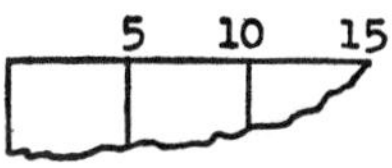

Figure 15

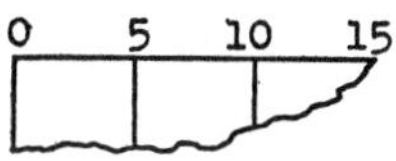

Figure 16

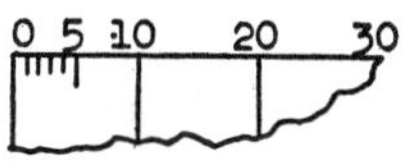

Figure 17

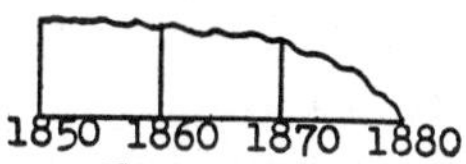

Figure 18

Figure 19

Figure 20

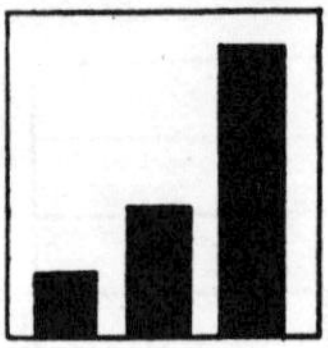

Figure 21

Figure 22

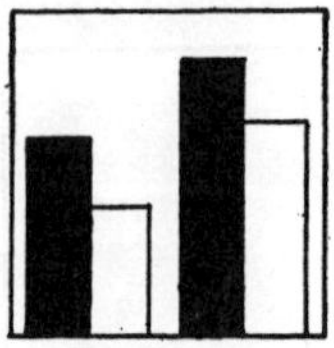

Figure 23

Figure 24

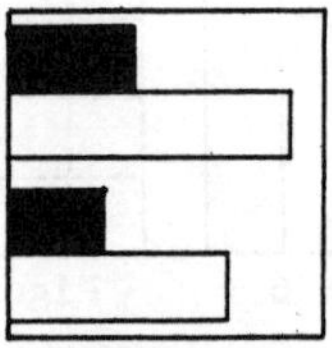

Figure 25

Figure 26

Figure 27

Figure 28

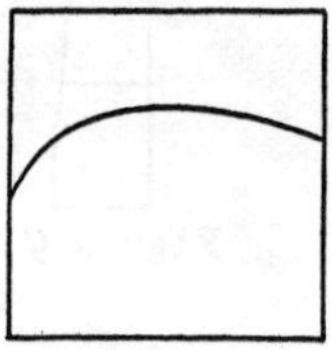

Figure 29

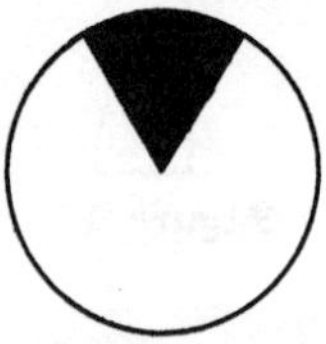

Figure 30

line and curved line graphs that were found. The purposes for which the different types of graphs were used are summarized in the first part of Table XVI. The same table shows the frequency with which each background and each scale was used and the types of graphs with which each was associated. It also contains a list of the items scaled and a description of the complexity of the graphs.

TABLE XVI

CHARACTERISTICS OF VARIOUS TYPES OF GRAPHS IN THE JUNIOR HIGH SCHOOL SOCIAL STUDIES MATERIALS EXAMINED

Characteristic	Type of Graph											
	Horizontal Bars			Vertical Bars			Sectored Circle			Line		
	Grades											
	7	8	9	7	8	9	7	8	9	7	8	9
Purpose												
To show comparison of												
Areas	3						1					
Budgets									1			
Causes			1									
Consumption	1											
Exports	2											
Immigration			1									1
Imports	2											
Coast lines	1											
Per cents	1						1	5				
Populations	2	3		1								
Production	8	2	2	1			19		18			
Uses												1
To show growth of												
Circulation of money												1
Empires				1								
Enrollments				1								
Imports				2								
Industrial establishments				1								
Industries				1								
Machines				1								
Manufacturing												1
Membership					1							
Population			1	3	1							2
Production	1	6	1	7	2							
Public high schools					1							
Railroads		1			1							2
Steam power											1	
Stockholders						1						
Tonnage of vessels				1							1	1
Trade	2				1							
Background												
Equal rectangles		1								2	1	
Equal squares											4	3
Vertical lines equally spaced	14	9	2									
Horizontal lines equally spaced				19	7							
Some subdivisions	3			3	1							
Plain	7	4	4			1						

TABLE XVI (Continued)

Characteristic	Type of Graph											
	Horizontal Bars			Vertical Bars			Sectored Circle			Line		
	Grades											
	7	8	9	7	8	9	7	8	9	7	8	9
Descriptive												
Black bars	17	13	4	18	5	1						
Black bars and white bars					1							
Black bars and white bars paired	2	1			1							
Black bars and shaded bars paired	2											
Divided bars			3	1								
Bars grouped in sets of two, three, or four	6	1		4	1							
White sectors							2		1			
Black sectors and white sectors							19	5				
Shaded sectors									18			
Broken line										2	3	1
Curved line											2	
Less than six bars	11	11	3	7								
Six to ten bars	8	3	3	10	4	1						
Eleven to twenty bars	2			2	1							
More than twenty bars			1		1							
Two sectors							18	5	18			
Four sectors							2					
Seven sectors							1					
Nine sectors									1			
One line										2	2	1
Two lines											1	1
Three lines												1
Four lines											1	
Five lines											1	
Scaling												
Horizontal scale												
Appears at bottom										2	5	1
Appears at top	14	9	3									2
Proceeds from left to right	14	9	1							2	5	1
Vertical scale												
Appears at left				19	3					2	5	1
Appears at right					4						1	
Proceeds from bottom to top				19	7					2	6	1
No scale	7	5	5									
Zero point is on the diagram	14	9	6	19	7	1				4	10	4
Zero point is labeled	6	9	2	1	5					2	4	3
Hundred per cent line is shown				1								

TABLE XVI (Continued)

Characteristic	Type of Graph: Horizontal Bars, Grades 7	8	9	Vertical Bars 7	8	9	Sectored Circle 7	8	9	Line 7	8	9
Total line is shown	1											
Numbers at left of bars	6	5	1									
Numbers at end of bars		3				1						
Sectors labeled in per cents							3	5	19			
One sector labeled in per cent							14					
Bar lengths increase												
Bottom to top	12	3	2									
Top to bottom	2	2	3									
Left to right				16	7	1						
Items listed or scaled												
Horizontally												
Barrels	2											
Board feet		1										
Dollars	4	6	3									
Gallons		1										
Miles	1	1										
Not indicated	1											
Per cents	4		1									
Persons	2	3	3									
Square miles	3											
Tons	4	2										
Years				19	7	1				2	5	1
Vertically												
Cars					1							
Causes			1									
Chests				1								
Classes of people	1											
Countries	15	5	1									
Dollars			1		1						1	
Food products	3											
Grains	1											
High schools					1							
Horsepower			1							1		
Industries				1								
Machines				1								
Miles					1							
Per cents			1	1								
Persons				3	2	1					2	
Pounds				1								
Square miles				1								
Tons				10	1							
Years	2	9	2									

All the graphs were well titled, and most of them were referred to in the exposition and occasionally interpreted.

The outcomes of the study as presented in the previous sixteen tables serve to emphasize the need for a clear understanding of the meanings and uses of certain elementary mathematical concepts. The mathematical activities, the problems, questions, and exercises, in the social studies that require the use of mathematical processes in formulating the solution or the answer, occur less frequently. Table XVII shows that the activities required involve a knowledge of the simpler uses of the compasses, ruler, and pencil; the ability to draw to scale; the ability to construct a simple bar graph; and the ability to apply a few of the simpler processes of calculation. With the exception of one problem involving the finding of the per cent one number is of another, all the activities were required as early as the seventh year. In some cases, as shown in the table, the processes were carried out in the social studies materials before they were required in the exercises. In other cases processes were carried out to clarify the explanation without requiring any further use by the pupil.

The two illustrations that follow show in detail the way in which mathematical processes were associated with the social studies materials:

1. But how do councils, commissioners, supervisors, or any other officials who have the duty of levying taxes decide how much they will collect from any person? By dividing the amount needed for public expenses by the total property valuation, a decimal is obtained, which is the rate of tax on the property of each individual or corporation. If the property valuation of a township is $10,000,000 and the amount to be raised for the property tax is $120,000, the quotient is .012. A person whose property is valued at $5,000 would then have to pay $60 in taxes to that community.[25]

2. We now count time from the birth of Christ. Any date before that we mark B. C.; any date after it, A. D. These last two letters stand for Latin words which mean "in the year of our Lord."

Each hundred years is called a century, and centuries are numbered just as you number years of your life. When you had finished your first year, you were one year old. Until then, you were in your first year. If you were twelve at your last birthday, you are now in your thirteenth year, which you will finish on your next birthday. So the first

[25] Hughes, R. O., *A Textbook in Citizenship,* p. 722. 1923.

TABLE XVII

MATHEMATICAL ACTIVITIES IN THE JUNIOR HIGH SCHOOL SOCIAL STUDIES MATERIALS EXAMINED

Activity	Grade 7	Grade 8	Grade 9
Required			
To add numbers	x		
To arrange names in time order	x		
To check an answer	x		
To construct a time line	x		
To determine century in which a given date lies	x		
To divide a line into a number of equal parts	x		
To draw a line	x		
To draw an arc	x		
To draw to scale	x		
To estimate the length of a line	x		
To estimate distance on a map	x		
To estimate heights by comparison	x		
To estimate population	x		
To find per cent one number is of another			x
To interpret or study a graph	x	x	x
To interpret or study a table	x	x	x
To make a graph	x		
To read a time line	x		
Illustrated			
Determining century in which a given date lies	x		
Equivalence of certain units of measure	x	x	
Finding interest on a loan		x	
Obtaining a tax rate and amount of tax			x
Scale drawing	x		

century A. D. was finished at the end of the year 100; the thirteenth at the close of the year 1300. 75 A. D. is a date in the first century after Christ; 1215 is a date in the thirteenth century; 490 B. C. is a date in the fifth century before Christ. In what century is each of the following dates: 550 B. C.; A. D. 1492; 1607; 1928? [26]

The supplementary activities listed in Table XVIII represent the activities required in the exercises in workbooks that have been published to accompany three of the textbooks that were examined. The activities, primarily associated with tables and graphs, include the making of graphs and tables and getting facts from graphs and tables already constructed.

[26] West, Ruth and West, Willis M., *The New World's Foundations in the Old*, p. 15. 1929.

TABLE XVIII

MATHEMATICAL ACTIVITIES IN WORKBOOKS TO ACCOMPANY JUNIOR HIGH SCHOOL SOCIAL STUDIES TEXTBOOKS

Activity	Workbook A	Workbook B	Workbook C
Answer questions about a graph	2		
Calculate the per cent one number is of another (method described in the workbook)			8
Compare two areas experimentally (method described in the workbook)	2		
Draw a bar graph to show			
Comparative size of areas or populations	1		2
Growth of cities	1		
Increase in automobiles		1	
Increase in export trade		1	
Increase in manufactures		1	
Increase in mineral products		1	
Increase in national wealth		1	
Increase in number of factories	1		
Increase in production of coal	4		
Increase in production of iron	4		
Increase in production of petroleum		1	
Increase in production of pig iron		1	
Increase in railroad mileage	1		
Percentages of people in principal economic groups (method described in the workbook)			1
Draw a circle	1		
Draw a line	5	2	1
Draw a line graph	2		
Draw a pictograph	2	1	2
Draw a table	1	2	1
Draw a time line (method described in the workbook)	5	5	5
Draw a vertical time line			1
Draw to scale	5	5	6
Estimate an area	1		
Fill in the columns of a table	13	9	8
Get statistical facts from a graph	3		
Get statistical facts from library sources	12	7	2
Get statistical facts from tables	1		1
List quantities in order of size	2		
Locate points on a time line	17	10	15
Select correct integers and fractions in multiple choice questions	10	10	2

The study of the mathematics encountered in the junior high school social studies materials was concluded with the examination of seventh and eighth grade textbooks in geography and history in common use. These books were examined for comparison purposes and were not included in the reference materials taken from the Denver syllabus. The examination made it evident that, for the most part, the type and frequency of the mathematics involved were not sufficiently different to justify a complete tabulation. The nature of the connections, graphs, and activities in one geography textbook [27] and in one history textbook [28] will be considered.

In the geography materials the discussion of the earth as a planet includes solar system facts involving mean distances from the sun in millions of miles and mean diameters of planets in miles; facts concerning the shape of the earth involving the turning of the earth round an imaginary central line called the axis, the polar diameter (distance through the earth from pole to pole) as 7,889.4 miles, and the path or orbit of the earth around the sun as an ellipse rather than as an exact circle; and facts about keeping time and standard time involving the notion of the rotation of the earth on its axis from west to east, the day as one revolution, the division of the day into 24 hours, the sunrise going round the earth in 24 hours or 360 degrees, the sunrise going 15 degrees in one hour, and the time required for a complete revolution around the sun as 365 days, 5 hours, 48 minutes, and 46 seconds. The discussion of latitude and longitude includes the explanation of the construction of the prime meridian and the equator; the dividing of circle into 360 degrees; the fact that a quarter of the way around the earth equals 90 degrees; parallels of latitude; the measuring of distances east and west in terms of degrees of longitude, and distances north and south of the equator in degrees of latitude; the obtaining of more exact locations by dividing each degree into 60 minutes and each minute into 60 seconds; the difference in length of degrees, 1/360 of the distance around the earth at the equator being much more than 1/360 of the distance around the earth near the pole; and tables of degrees of longitude in statute miles to thousandths.

The Homolosine projection as an equal area projection in map making whereby a square inch anywhere on the map represents the same number of square miles of the earth's surface as any other square inch; scales of miles; heights of land and altitudes above and below sea level; elevations of areas; differences of pres-

[27] Smith, J. Russel, *Human Geography, Book Two.* 1930.

[28] Tryon, Rolla M. and Lingley, Charles R., *The American People and Nation.* 1929.

sure as recorded by the barometer; air pressure at sea level on the average as fifteen pounds per square inch; the pressure of the air when the barometer is high as sometimes fifteen and a half pounds to the square inch, or 2,232 pounds to the square foot; graphs of inches of rainfall on the average for each month; measurement of rainfall, as 14 inches, 27½ inches, 32.5 inches, or 25.04 inches; measurement of temperature with the thermometer, as from

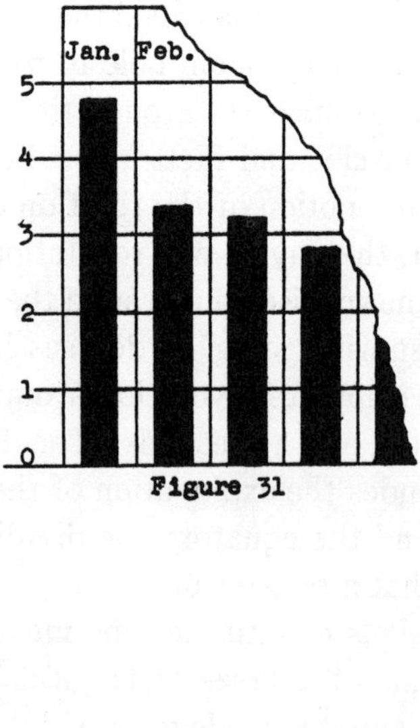

Figure 31

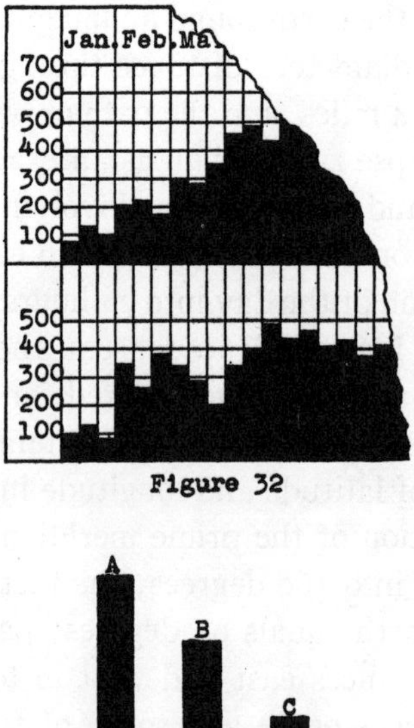

Figure 32

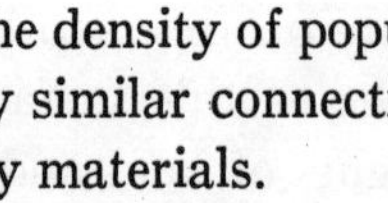

Figure 33

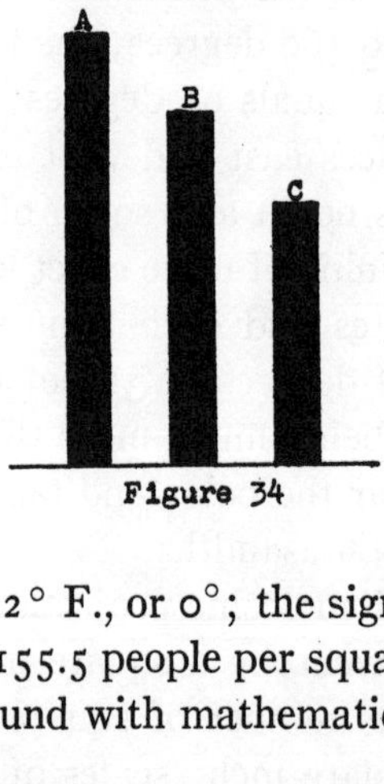

Figure 34

twenty to forty degrees below zero, 22° F., or 0°; the significance of 32°; the density of population as 155.5 people per square mile; and many similar connections are found with mathematics in the geography materials.

Three tables not associated with graphs, one broken line graph, one horizontal graph, and thirty-four vertical bar graphs occurred in the geography textbook. The graphs were used primarily to show differences in rainfall, in production of various commodities, and in populations. Figure 31 on page 40 illustrates the type most commonly used. The type in Figure 34 frequently occurred, accompanied by a table to explain the use and extent of A, B, etc. In the history textbook seventeen tables and six graphs appeared. The type of graph in Figure 33, unsatisfactory for purposes of comparison, occurred in both textbooks.

The mathematical activities required in the geography and in the history textbook are listed in Table XIX.

TABLE XIX

Mathematical Activities in the Geography and the History Textbook Examined

Activity	Textbook G	Textbook H
Draw a bar diagram to illustrate an increase		18
Draw a graph to show comparison	8	
Make a table		2
Prepare a time line		1
Tell what a diagram shows	5	
Draw a 4,000-mile circle around northern New Zealand, and tell what proportion of the area is land	1	
Determine a required length on a map, using the map scale	2	
Determine the size of the Great Plains in square miles from the data on a given page. (Data: If it were stretched out like a long, straight path or highway ten miles wide, it would go nearly four times around the earth)		1
Find on the world map the place where the other end of the tunnel would be. It is 180° east longitude and 51½° latitude. Find the place that is on the other side of the earth from Pittsburgh (longitude 100° east and latitude 41° south; the place that is opposite Cape Town (34° north and 162° west)	1	
If you should go as far south of the equator as Paris is north of the equator, you would be at the point from which a ship could sail east or west in a straight line entirely around the world, except for one bit of land. What land is it? How wide in degrees of longitude? What part of the distance around the world is this?	1	
How many degrees of longitude across the widest part of the Sahara? Across the widest part of the United States? How far is it from Cairo to Basia? From New York to Chicago?	1	
In one hour it (the sunrise) goes 15 degrees. How long does it take to go one degree?	1	
A ship from Galveston once unloaded at Havre 28,000 bales of cotton of 500 pounds each. What would that be worth now?	1	
Land that was given away in 1860 and 1870 is now worth $100 an acre, and in some places as much as $200 or $300. What does it cost to get a 160-acre farm now?	1	
If wild rabbits double their number every three months, how many offspring would one pair of rabbits have in five years?	1	

SUMMARY

It is possible at this point to summarize and compare the mathematics found in the junior high school social studies materials examined with stated objectives in mathematics for the junior high school years. The summary of the findings will be formulated in comparison with one such list of objectives.[29] The comparison will be limited to the objectives in arithmetic, since it appears evident that the mathematics encountered thus far in the social studies materials does not justify a consideration of the objectives in algebra or in geometry.

In the field of whole numbers it is necessary to consider objectives in counting; reading and writing numbers; the degree of accuracy in writing results; addition, subtraction, multiplication, and division; and the use of short methods. So far as the materials of instruction examined in the social studies are concerned, it is necessary to understand the meaning of counting and to realize that the majority of the numbers in use in the social studies are the results of counting. Reading numbers is satisfied by the general objective that "as far as the needs of most people are concerned, the reading of numbers may well be limited to billions." The construction of graphs and the requirement to make and fill in tables necessitate ability to write numbers, but in general the numbers are not large. Pupils should understand what is meant by the degree of accuracy to which numbers are written, but they will find more occasion for interpreting such numbers in their reading than for writing them or for exercising judgment as to the degree of accuracy to which numbers should be written. They will find very little actual need for performing the operations of addition, subtraction, multiplication, and division with whole numbers, but they will find these processes illustrated and the concepts appearing frequently in the exposition.

Under the objectives with common fractions it is stated that "the common fractions which we still need to use are halves, thirds, fourths, eighths, tenths, twelfths, sixteenths and in rare cases fifths, sixths, sevenths, and thirty-seconds." These are the fractions that are used most frequently in the social studies materials,

[29] Smith, D. E. and Reeve, W. D., *The Teaching of Junior High School Mathematics,* pp. 91-130. 1927.

but the pupil will find an occasional use of several others, and his mathematical training must provide him with the means of interpreting them. The processes of adding, subtracting, multiplying, dividing, and reducing fractions do not occur.

The objectives with decimals are included in the statement that "when the pupil understands the principle which underlies the method of expressing decimals, he will have no trouble in reading or writing them." The decimals that occurred were small in number and limited in range. No skills or abilities with decimals were found necessary except the ability to read and understand them.

The concept of per cent, usually considered in connection with decimals, is one of the most frequently used mathematical terms and its meaning should be clearly understood. The pupil will find it necessary to read and understand simple expressions, such as ten per cent and 54 per cent. He will find it equally necessary to comprehend fifty-one and four tenths per cent, 10.04 per cent, 150 per cent, 2,465 per cent, and one half per cent. One instance was found in which it was necessary to determine what per cent one number is of another.

Units used in measuring length, surface, capacity, arc, time, and temperature appeared frequently in the social studies. An understanding of the equivalence of measures and an appreciation of the applications of different measures in a variety of different settings were demanded. Compound denominate numbers occurred in the simple forms 5 ft. 3 in. and 36° 30′. A knowledge of scale drawing is necessary both in reading and in making drawings to scale.

Aside from a few commercial and geometric terms usually associated with the study of arithmetic, the remaining abilities required are associated with tables and graphs and, from the standpoint of the frequency of their use in the social studies materials, these abilities should receive careful attention. The ability to interpret graphs and the ability to construct graphs are required in the social studies materials examined early in the seventh year. "The pupil should know that facts may be represented by tabular, graphic, or formula methods. The table is necessary, because it gives all the details; the graph, because it pictures the facts more vividly than the table." These facts about the table and graph the pupil will need to appreciate. He should realize also that different situations are pictured best by the proper selection from various types of graphs—the circle graph, the bar graph, the line graph, and the pictograph.

The kind and number of uses of mathematics has been treated as only one aspect of the relation between mathematics and the social studies. The social studies, like mathematics, emphasize meanings and appreciations in assisting pupils to acquire proper attitudes and useful factual knowledge. The connections in which mathematics occurs in the social studies materials afford a basis for associating the learning of mathematics with real situations and for developing an appreciation of the power of mathematics as an instrument in social situations and in social progress. Only a very small minority of the pupils will gain an awareness of the relations between the artificial separations of subject matter in their learning unless a consistently deliberate effort is made by all subject matter representatives to create such an awareness through assisting pupils to gain a complete perspective of the chief ends of all learning.

CHAPTER III

CONCEPTS AND PROCESSES IN ARITHMETIC AND JUNIOR HIGH SCHOOL MATHEMATICS TEXTBOOKS

INTRODUCTION

THUS far this study has been concerned with the actual mathematics that occurs on the printed pages of the junior high school social studies materials. It is equally important to know the extent to which provision is made to assist pupils to understand the mathematics that occurs in these materials. Such provision is dependent upon a number of variable factors including the abilities and the interests of the pupils, the preparation and teaching skill of the teachers, and the materials of instruction in mathematics. In any given situation, undoubtedly the content of the mathematics textbook is most nearly constant. For this reason the concepts and processes explained in arithmetic and in junior high school mathematics textbooks have been selected as representative of the probable training in mathematics that pupils receive preceding and in conjunction with their prescribed courses in the social studies at the junior high school level. The determination of these concepts and processes is indicative of the provision that is made for training in mathematics as revealed by an analysis of the textbooks. This analysis in many cases may have little relation to the limits of instruction or to the results of the training.

SELECTION OF THE TEXTBOOKS

Eighteen textbooks in arithmetic in use prior to the seventh year and nine textbooks in use in the seventh, eighth, and ninth years were selected. The eighteen textbooks in arithmetic were from eight different series. All were recent editions published between 1928 and 1934. The arithmetics are listed in Appendix II and the junior high school books in Appendix III.

RESULTS OF THE ANALYSIS OF THE TEXTBOOKS

In the arithmetic textbooks 212 defined terms were found, and in the junior high school mathematics textbooks there were 672 defined terms. In addition, 131 units of measure were in use.

A number of symbols and abbreviations occurred, including the signs of addition, subtraction, multiplication, division, and equality; the brackets; the parentheses; the symbols for number (♯), cent (¢), dollar ($), degree (°); second (″), minute (′), inch (″), foot (′), decimal point (.), caret (∧), colon (:), period (.), bar (—), at (@); and the abbreviations for before noon (A.M.), after noon (P.M.), and for units of measure in common use.

In Table XX the topics that were explained or illustrated in the textbooks are listed by grades, with the number of different series in which each topic was explained or illustrated. A study of the table shows that there is considerable agreement about the topics included at the various grade levels. Graphs, ratio, and tables are the only major topics in the arithmetics that are not included in all the textbooks. In the higher grades there is some variation in

TABLE XX

TOPICS EXPLAINED IN ARITHMETIC AND JUNIOR HIGH SCHOOL MATHEMATICS TEXTBOOKS EXAMINED

Topic	Number of Series	Topic	Number of Series
	First Six Grades		
Addition	8	Multiplication	8
Approximation	8	Per cent	8
Area	8	Problem solving	8
Average	8	Rate	8
Business uses	8	Ratio	5
Checking	8	Roman numerals	8
Decimals	8	Scale drawing	8
Division	8	Subtraction	8
Fractions	8	Tables	7
Geometric terms	8	Units of measure	8
Graphs	5	Volume	8
Measurement	8	Whole numbers	8
	Seventh Grade		
Approximation	3	Business forms	3
Average	3	Commission	3
Budget	3	Decimals	3

TABLE XX (Continued)

Topic	Number of Series	Topic	Number of Series
Denominate numbers	3	Latitude and longitude	2
Discount	3	Measurements (lengths, areas, volumes)	3
Formulas	3	Measuring instruments	3
Fractions	3	Metric system	3
Fundamental operations	3	Percentage	3
Geometric constructions	3	Problem solving	3
Geometric figures (angles, lines, polygons, triangles)	3	Profit and loss	3
Geometric relationships	3	Ratios	3
Graphs	3	Scale drawing	3
Insurance	3	Symmetry	1
Interest	3	Thrift and banking	3
		Eighth Grade	
Accounts	3	Measuring instruments	3
Algebraic expressions	3	Measurements (lengths, areas, volumes)	3
Algebraic language	3	Metric system	3
Approximation	2	Mortgages	3
Banking	3	Percentage	3
Bonds	3	Problem solving (arithmetical and algebraic)	3
Budget	2	Profit and loss	3
Congruence	2	Promissory notes	3
Denominate numbers	3	Proportion	3
Discount	3	Pythagorean principle	3
Equations	3	Ratio	2
Formulas	3	Roots and powers	3
Fundamental operations (arithmetic, algebra)	3	Scale drawing	3
Geometric constructions	3	Signed numbers	3
Geometric figures	3	Similarity	3
Geometric relationships	3	Stocks	3
Graphs	3	Symmetry	2
Indirect measurement	3	Taxation	3
Insurance	3	Trigonometric ratios	3
Interest	3		
		Ninth Grade	
Algebraic expressions	3	Graphs	3
Algebraic language	3	Logarithms	1
Approximation	2	Numerical trigonometry	2
Equations (through quadratics)	3	Problem solving	3
Factoring	3	Radicals and powers	3
Formulas	3	Signed numbers	3
Fractions	3	Slide rule	1
Fundamental operations	3	Variation and dependence	2

the consideration of symmetry, congruence, logarithms, the slide rule, variation, and a few other topics. The examination of more textbooks undoubtedly would tend to show other variations, but the close agreement in the textbooks examined appeared not to justify further examination.

Although the topics in the various textbooks are similar, the defined concepts that are associated with these topics vary extensively. Slightly less than one fourth of the defined terms in the arithmetics were defined in all the series examined. More than half the terms were defined in less than half the series. A part of this variation can be accounted for by the fact that terms defined in some series are used without definition in others and that slightly different meanings often are given to the same terms and slightly different terms are used to convey similar meanings. The variation of most significance, however, arises from the different treatments of the same topics. In the junior high school mathematics textbooks the situation is similar. Less than one out of five defined terms was defined in all the series, with large differences on each grade level. It seems evident that the mere inclusion of a topic gives little indication of the treatment or the learning involved.

SUMMARY

Defined terms have been emphasized in the discussion of the content of the mathematics textbooks because the social studies constantly require interpretations in terms of the meanings of concepts. The extent to which the mathematics textbooks examined provide a means for pupils to become acquainted with the mathematical concepts and processes found in the social studies materials can be studied by comparing the findings in Chapter II with the topics in Table XX and the terms associated with these topics.

The extensive use of number in the social studies materials necessitates a complete understanding of the meaning of number. Preparation for this understanding begins in the informal work in arithmetic in the early experiences of the pupil and in the first grades of the elementary school. This preparation is continued in textbooks in arithmetic and in other mathematics in the later grades. The specific number demands made by the junior high school social studies materials examined include the use of ordinal numbers expressed in Arabic notation, in Roman notation, and in words; cardinal numbers expressed in Arabic notation, in words, and in a combination of the two, ranging in size from zero to billions; fractions with denominators ranging from two to a million; and decimals written to one, two, and three decimal places. The meanings of all these number concepts are treated in the arithmetic textbooks adapted for the first six grades.

The numbers appearing in the social studies materials are the results of counting and measuring. All measurement is approximate, and many of the results of counting are expressed approximately or are estimates. The idea of the approximate number is treated in every series of arithmetic textbooks examined. The definition of estimate appears in three of the eight series and the definition of round number in one. The notion of approximation is more often limited to writing a number correct to a nearest decimal place than to expressing the possible degree of accuracy in measurement or computation or to expressing estimates. More attention is given to the latter type of approximation in the junior high school mathematics textbooks, but the use in the social studies materials is as frequent in the seventh year as in the later years.

Not all the units of measure in use in the social studies materials appeared in the mathematics textbooks examined. This is particularly true of units not in common use or not within the experiences of the pupils. Such units include the dram, the guinea, the peso, and others. It is quite unnecessary and probably not desirable that all these units should appear in the mathematics textbooks. But mathematics textbooks can assist in meeting these demands of the social studies as they occur by providing an appreciation of the use of different systems of measurement and a knowledge of sources where the meanings and equivalents of uncommon units can be found.

The mathematical symbols found in the social studies materials are not unlike those found in the mathematics textbooks.

The frequencies with which a number of terms in the mathematical vocabulary in Table XI were found seem to indicate that some of these terms ought to be given more attention in the mathematics textbooks than at present appears to be common practice. In spite of their frequent appearance in the early social studies, very little attention is given to defining such geometric concepts as circle, radius, center, arc, pyramid, hemisphere, diagonal, or cylinder sufficiently early in mathematics. Bar graph and circle graph are defined only in a few of the arithmetics below the seventh year, although the terms are frequently used in materials in the social studies early in the seventh year. Proportion, average, and rate are used more frequently than their treatment in the arithmetics might suggest.

One fourth of the series of arithmetic textbooks that were examined provided no treatment of graphs below the seventh year, two series consider only the broken line graph, and no series considers all the types of graphs appearing in the social studies materials. Graphs appear early in the seventh year in the materials in the fusion course in the social studies and in the geography materials. Among these graphs are horizontal and vertical bar graphs, circle graphs, and broken line graphs. It is essential that provision be made for a consistently earlier introduction of graphs, if pupils are to comprehend the use of graphs in the social studies materials examined in this study. Adequate preparation is made in the junior high school mathematics textbooks for the comprehension of the statistical graphs appearing in the social studies in the eighth and ninth years.

The mathematical activities required in the social studies, aside from the construction of graphs, are adequately provided for in the arithmetic textbooks. It is possible that both mathematics and the social studies might profit if mathematics textbooks made more use of the connections in which mathematics is used in the social studies materials.

CHAPTER IV

MATHEMATICS IN PRESCRIBED INTRODUCTORY COURSES IN SOCIAL STUDIES AT THE COLLEGE LEVEL

NATURE OF THE PRESCRIBED COURSES

An examination of the catalogues of a number of colleges and universities tends to show that there is little uniformity in the extent or the nature of the social studies courses required for graduation. Many of the institutions state a definite requirement in terms of units or points of credit, and then permit the student to fulfill the requirement by making his own selection from catalogued groups of courses in history, economics, sociology, and other divisions of the social studies. Other institutions prescribe definite courses for all their students during certain periods of their college course. Some of the courses are characterized by the traditional subject matter divisions of the social studies, whereas others are organized on a fusion plan to include materials from all the social studies as the organization demanded. Fusion has resulted in an extensive variation in the organization of the social studies materials at the college level, but to a lesser degree in the materials of instruction that are used.

A quotation from a recent university catalogue [1] will serve to state the aim and nature of one prescribed fusion course in the social studies for college freshmen. "The aim of this course is to inform the student of the more important factors in his contemporary society and especially to increase his understanding of the economic, political, and intellectual background of the present day. The European foundations of contemporary American culture are studied, and the economic and intellectual development of the United States is given particular consideration. This course prepares the student for the intensive study of economic and political problems in the United States."

[1] *Announcement of Columbia College,* 1934.

SELECTION OF THE SOCIAL STUDIES MATERIALS

The chief purpose of analyzing the materials of instruction used in prescribed social studies courses at the college level to determine the mathematics involved is to determine the extent of the additional mathematical demands made by the social studies beyond the demands made in the early years of secondary education. The prescribed fusion courses in the social studies are well suited to this purpose because of the wide range of the materials in use in these courses. The reference materials were selected from the required readings in the prescribed social studies course described in the preceding paragraph. The complete list of books examined is given in Appendix IV.

RESULTS OF THE ANALYSIS OF THE MATERIALS

A study of the numbers that occurred in the materials for this period showed that ordinal numbers continue to occur in similar forms and with uses similar to those found in the materials for the junior high school grades. In the history references the use of Roman numerals was somewhat more frequent. But from the

TABLE XXI

FREQUENCY OF OCCURRENCE OF LARGE CARDINAL NUMBERS IN THE PRESCRIBED COLLEGE SOCIAL STUDIES MATERIALS EXAMINED

Cardinal Numbers	Total Frequency in All Books	Total Number of Books
In Figures		
Ten digits	48	6
Eleven digits	18	1
Twelve digits	8	1
Twenty-one digits	1	1
Sixty-seven digits	1	1
Subtotal	76	6
In Words		
Ten digits	33	5
Eleven digits	16	3
Thirteen digits	4	2
Twenty-one digits	1	1
Subtotal	54	5
Total	130	8

standpoint of additional training in mathematics, it is quite unnecessary to consider ordinal numbers further.

What is true of ordinal numbers is equally true of the common integers among the cardinal numbers. The occurrence of the larger integers, those of ten places or more, are summarized in Table XXI. Numbers of this size were found in seven of the nine

TABLE XXII

FREQUENCY OF OCCURRENCE OF FRACTIONS IN THE PRESCRIBED COLLEGE SOCIAL STUDIES MATERIALS EXAMINED

Fraction	Total Frequency in All Books	Total Number of Books
Nine tenths	7	3
Seven eighths	1	1
Five sixths	2	1
Four fifths	9	5
Seven ninths	1	1
Three fourths, three quarters, 3/4	38	6
Five sevenths	1	1
Seven tenths	4	3
Two thirds	42	8
Five eighths	4	3
Three fifths	6	2
One half, half, 1/2	257	9
5/12	1	1
Two fifths	9	4
Three eighths	1	1
One third, third, 1/3, 4/12	57	7
Three tenths	1	1
One fourth, fourth, one quarter, quarter, 1/4, 3/12	68	8
One fifth, fifth	18	5
One sixth, sixth	14	5
One seventh, seventh	3	2
One eighth, eighth	6	4
One tenth	8	3
One eleventh	1	1
One twelfth	3	2
One sixteenth	1	1
One twentieth, twentieth	3	3
One hundredth, hundredth	3	2
Thousandth	1	1
760 millionths	1	1
390 millionths	1	1
One ten-thousandth	1	1
One millionth	1	1
10^{-10}	1	1
Total	575	9

books examined in expressing the results of counting people, money, acres, nerve cells, horsepower, and years. The largest numbers written out contained twenty-one digits. The use of trillion and quintillion was introduced. The latter of these concepts arose in considering the depreciation of the German mark. Symbolic notation was illustrated by the use of 10^{66}.

The fractions, decimals, and units of measure used were listed and are summarized in Tables XXII, XXIII, and XXIV. A study of these tables leads to the conclusion that the use of the concepts of number and of measurement is largely a repetition of the concepts that appeared in the materials in the earlier grades. Differences in the nature of the fractions appearing are represented by

TABLE XXIII

FREQUENCY OF OCCURRENCE OF DECIMALS IN THE PRESCRIBED COLLEGE SOCIAL STUDIES MATERIALS EXAMINED

Decimal	Total Frequency in All Books	Total Number of Books
Tenths		
Significant figures		
One	5	2
Two	29	4
Three	104	5
Four	22	3
Five	4	2
Subtotal	164	5
Hundredths		
Significant figures		
One	1	1
Two	3	1
Three	39	4
Four	15	4
Five	1	1
Subtotal	59	5
Ten Thousandths		
Significant figures		
Five	1	1
Subtotal	1	1
Millionths		
Significant figures		
Four	1	1
Subtotal	1	1
Total	225	5

TABLE XXIV

FREQUENCY OF OCCURRENCE OF UNITS OF MEASURE IN THE PRESCRIBED COLLEGE SOCIAL STUDIES MATERIALS EXAMINED

Unit	F[a]	N[b]	Unit	F[a]	N[b]
Acre	75	4	Lira	4	2
Angstrom unit	1	1	Livre	2	1
Barrel	11	2	Mark	31	3
Bushel	11	3	Meter	2	2
Cent	30	5	Metric-ton	1	1
Century	1,175	9	Mile	139	7
Chervonetz	1	1	Millimeter	1	1
Crown	2	1	Minute (arc)	3	1
Cubic millimeter	1	1	Minute (time)	20	5
Day	389	9	Month	130	9
Decade	56	9	Ounce	2	1
Degree (angle, arc)	375	3	Pence	28	3
Degree (temperature)	9	2	Penny	5	3
Dollar	441	8	Pound (sterling)	226	6
Dozen	26	8	Pound (weight)	44	4
Drachma	1	1	Quart	1	1
Eon	1	1	Rod	1	1
Foot	12	4	Ruble	3	1
Franc	27	3	Score	7	3
Farthing	3	1	Second (time)	14	3
Gallon	3	2	Shilling	95	4
Grain	2	1	Sou	2	1
Guinea	3	1	Square meter	3	1
Hectare	1	1	Square mile	72	4
Hogshead	4	1	Tael	5	1
Horsepower	11	4	Ton	109	5
Hour	115	8	Yard	5	3
Hundredweight	5	3	Year	1,190	9
Inch	10	4	Week	119	8
Kilogram	2	1	Yen	1	1
Kilometer	6	2	Zloty	3	1
League	2	1			
Light year	2	2	Total	5,081	9

[a] Column F = Total frequency of the unit in all books examined.
[b] Column N = Number of books in which the unit occurred.

760 millionths, one ten-thousandth, and 10^{-10} The increased frequency of occurrence of decimals is accounted for by their use in statistical tables; the increased range, by the use of measurements of the nature of 1.2977 cents and .003937 inch. An occasional use of a unit from the metric system and a few additional units from foreign monetary systems account for the chief differences in the units listed in Tables IX and XXIV.

A few additional number concepts occurred. One radical expression was encountered. Exponents appeared twice. For convenience in expressing a very small number 10^{-10} was used; 10^{66} as an example of symbolism. Positive and negative numbers were found in a discussion of correlation and for the purpose of showing direction of change in balances. Various ways of expressing a ratio occurred. In comparing the strength of navies, 5:5:3; 5:5:3:1.67:1.67; and 5:5:3:1.75:1.75 were used. Other uses included "the proportion of the electors to the population was 1 to 32" and "the ratio of gold and silver liabilities decreased from 54.3 to 9.4." Expressions of such types as five feet six inches tall, the product of 48 times 34, and 300° F., continued to occur. Index numbers appeared once.

Thirty-one tables were distributed in five different books. The three graphs illustrated in Figures 35, 36, and 37 appeared. No mathematical activities were required.

Although there was very little significant growth in the frequency or use of number, there was a marked increase in the number and character of the mathematical terms in use. The mathematical terms that occurred are listed in Table XXV, but the significance of these terms can be understood only from a consideration of their connection with the social studies materials.

The statements of a number of mathematical facts and principles like those that follow make use of mathematical terms.

1. The whole is equal to the sum of its parts.
2. Between any two points there can be drawn a unique (i.e., straight) line.
3. A straight line is the shortest distance between two points.
4. Through a given point outside a straight line, there can be drawn but one parallel to that line.
5. About any point as a center circles can be drawn.
6. A circumference is the locus of all points touched by the end of a string or stick revolving around the center.
7. An angle in a semicircle is a right angle.
8. A triangle is known when one side and the adjacent angles are given.
9. The position of a point on a plane is determined when its relation to two known points has been established.
10. The forces must be reciprocally as the squares of their distances from their centres.

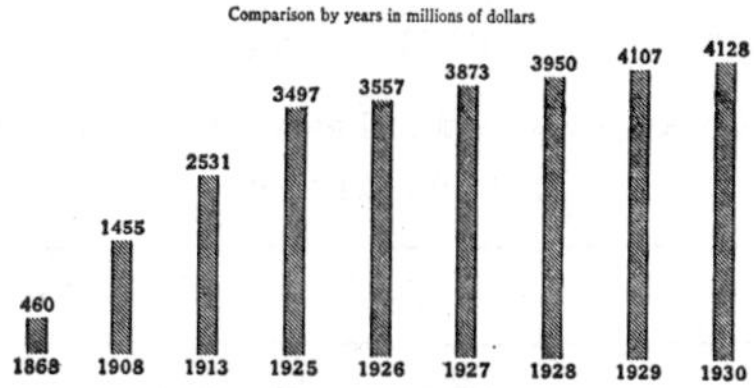

FIGURE 35. World Expenditures for Armaments
*From Langsam, p. 211.**

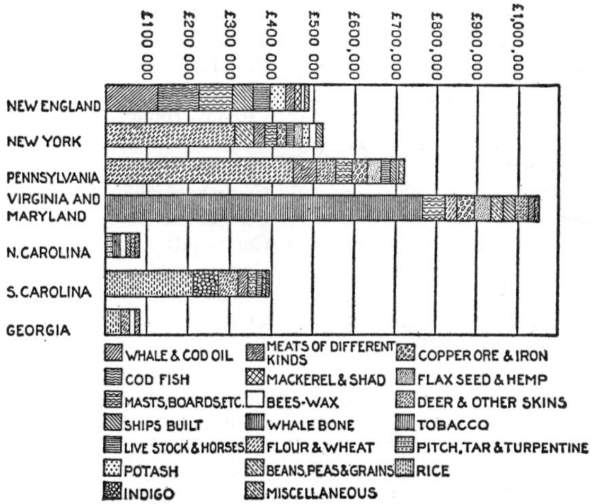

FIGURE 36. Annual Average of Colonial Exports, 1763-1773
From Carman, p. 161.†

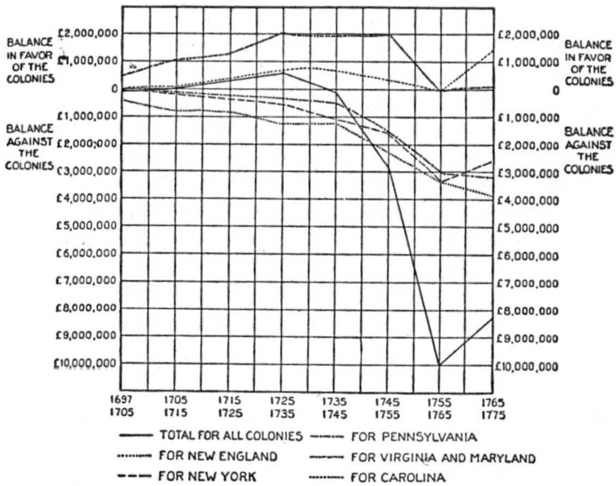

FIGURE 37. Balance of Trade Between the Colonies and Great Britain by Decades, 1697-1775
From Carman, p. 161.†

* Langsam, Walter C., *The World Since 1914*. The Macmillan Company, 1933. By permission of The Macmillan Company.

† Carman, Harry J., *Social and Economic History of the United States*, Vol. I. D. C. Heath and Company, 1930.

TABLE XXV

MATHEMATICAL VOCABULARY IN THE PRESCRIBED COLLEGE SOCIAL STUDIES MATERIALS EXAMINED

Term	F[a]	N[b]	Term	F	N
Absolute number	1	1	Chronometer	2	2
Abstract number	4	2	Chronoscope	2	1
Accelerate	9	3	Cipher (noun)	2	2
Acceleration	7	2	Cipher (verb)	1	1
Adjacent angles	1	1	Circle	36	5
Add	102	9	Circular	7	2
Algebra	13	3	Circumference	8	4
Algebraic	6	4	Circumscribing	1	1
Analysis	11	3	Coefficient of correlation	2	1
Analytical geometry	5	2	Coincide	1	1
Analytical method	3	2	Commensurable	1	1
Analytic geometry	1	1	Compass	3	2
Angle	10	5	Computation	1	1
Angular	4	3	Compute	1	1
Annuity	35	3	Concentric(ally)	2	1
Apex	1	1	Cone	2	1
Applied mathematics	2	2	Conics	1	1
Approximate	86	9	Conic curves	1	1
Arabian mathematics	1	1	Conic sections	3	3
Arabic notation	1	1	Continuous	3	3
Arc	3	2	Converge	1	1
Area	151	9	Coordinates	5	2
Arithmetic	8	3	Correlation	4	1
Arithmetical	2	2	Count	17	5
Arithmetician	1	1	Cube (power)	1	1
Assumption	3	2	Curve	21	5
Astrolabe	1	1	Cylinder	30	3
Average	45	8	Date	63	8
Axiom	8	1	Decimal	1	1
Axis	4	2	Decimal system	2	2
Balance	10	3	Deflection	1	1
Barometer	5	2	Degree (equation)	2	1
Binomial theorem	2	2	Degree of scatter	1	1
Bond	12	5	Demonstration	2	2
Calculate	21	4	Determined	1	1
Calculation	20	3	Deviation	4	1
Calculus	13	3	Diameter	5	3
Calendar	7	3	Difference	9	4
Calipers	1	1	Differential calculus	3	2
Cardinal numbers	1	1	Differential equations	2	1
Cartesian geometry	1	1	Digit	3	1
Center, centre	18	4	Dimension	3	2
Central point	1	1	Direction	5	3
Central tendency	2	1	Discounting	1	1
Chance distribution	2	1	Distribution	3	1

[a] Column F = Total frequency of the term in all books examined.
[b] Column N = Number of books in which the term occurred.

TABLE XXV (Continued)

Term	F	N	Term	F	N
Divide(d), division	78	9	Independent variable	1	1
Dividend	17	5	Index	1	1
Duodecimal	1	1	Infinitesimal	2	1
Ellipse	5	2	Infinitesimal calculus	1	1
Elliptical orbits	1	1	Infinity, infinite	63	4
Elliptic functions	1	1	Inscribed	1	1
Enclosed curve	1	1	Interest	39	7
End	6	3	Interferometer	1	1
Epicycle	2	2	Intersect	2	1
Epicycloid	1	1	Inversely proportional	2	1
Equal	56	9	Laterally	1	1
Equate	1	1	Latitude	6	4
Equation	13	2	Line	67	9
Equidistant	1	1	Linear	2	2
Equivalent	7	3	Locus	2	2
Error	1	1	Logarithm	2	1
Estimate	73	9	Longitude	14	4
Euclidean geometry	1	1	Mathematical(ly)	121	6
Euclidean system	1	1	Mathematician	30	6
Exhaustions	2	1	Mathematico-mechanical	1	1
Face	1	1	Mathematico-physical	1	1
Figure (compute)	5	3	Mathematics	146	6
Figure (geometry)	16	4	Mean	1	1
Figure (number)	46	7	Measure	54	8
Finite	10	2	Measurement	40	7
Fixed	5	3	Mechanical-mathematical	1	1
Fluxions	2	2	Median	1	1
Focus	1	1	Micrometer	1	1
Formula	21	5	Milliard	6	3
Fraction	9	4	Minus	3	1
Frequency	2	1	Multiply, multiplication	22	7
Function	6	2	Mode	2	1
Geometer	5	3	Negative	1	1
Geometric	3	2	Net gain	1	1
Geometrical	23	3	Non-Euclidean geometry	1	1
Geometrician	1	1	Normal distribution	1	1
Geometrise	1	1	Numerical	5	4
Geometry	41	6	Number system	2	1
Graph	2	2	Oblong	2	2
Graphic(ally)	5	2	Odd number	2	2
Hemisphere	10	5	Orb	7	2
Horizontal(ly)	4	3	Orbit	12	2
Hypergeometric series	1	1	Ordinal numbers	1	1
Hypothesis	5	3	Origin	1	1
Imaginaries	1	1	Oval	2	1
Imaginary line	1	1	Parabola	1	1
Incommensurable	2	2	Parallel	5	3
Increment	4	2	Parallel lines	2	2

TABLE XXV (Continued)

Term	F	N	Term	F	N
Parallel planes	1	1	Semicircle	2	1
Par value	3	2	Series, serial	7	2
Per cent	304	8	Sextant	2	2
Percentage	13	5	Side	1	1
Perfect circle	2	1	Slide rule	1	1
Perfect line	1	1	Slope	2	2
Perfect number	2	2	Skewness	1	1
Perpendicular lines	7	2	Solids	2	2
Physico-mathematical	2	1	Solid figures	1	1
Pictograph	2	1	Solid geometry	1	1
Plane	7	3	Solution	3	2
Plotting	2	2	Spatial relations	6	2
Plus	21	4	Sphere	71	7
Point	22	5	Sphericity	2	2
Positive	5	2	Spiral	2	2
Postulate	11	2	Square (geometry)	14	5
Principal	7	3	Square (power)	4	2
Prism	1	1	Square root	2	2
Probability	2	1	Statistical	8	3
Product	1	1	Statistician	3	1
Projection	2	2	Statistics	11	5
Proof	2	2	Straight line	6	3
Proportion	57	8	Spherical	2	2
Proposition	2	2	Stock	18	3
Prove	4	2	Stockholder	1	1
Pyramid	11	3	Subdivide	7	3
Quadrilateral	1	1	Subtract	3	2
Quadrant	3	3	Sum	67	9
Quadrature of curves	1	1	Summing	2	1
Quantitative	9	4	Superposition	1	1
Quotient	1	1	Symmetry	5	3
Radius	2	2	Symmetrical	1	1
Range	5	1	System of equations	1	1
Rate	86	9	System of notation	7	2
Ratio	14	6	System of numeration	3	1
Reciprocally	2	2	Table	6	4
Reckon	2	2	Tangent	1	1
Rectangular	1	1	Theorem	8	3
Rectilinear	1	1	Theory of proportion	1	1
Regular geometric faces	1	1	Three-dimensional	4	1
Regular bodies	2	1	Time notation	1	1
Relative numbers	1	1	Times	53	9
Relativity	3	1	Thermometer	7	2
Right angle	3	2	Third dimension	1	1
Round number	4	3	Transformations	1	1
Scale	45	4	Transverse	2	1
Sector	2	1	Triangle	7	4
Selling price	1	1	Trigonometrical	1	1

TABLE XXV (Concluded)

Term	F	N	Term	F	N
Trigonometry	3	1	Vertical axis	1	1
Two-dimensional	3	1	Volume	17	7
Unit(y)	35	6	Variation	6	3
Variable	3	3	Zero	3	2
Vary	13	5			
Vertical(ly)	5	3	Total	3,255	9

11. The acceleration of all the planets toward their centre was inversely proportional to the squares of their distances from it.
12. There is a fixed relation between the cubes of the distances of the planets from the sun and the squares of the times of their revolutions.
13. The straight line joining the planet to the sun sweeps out equal areas in any two equal intervals of time.
14. There is a law of the uniform downward acceleration of bodies falling to earth, of about thirty-two feet in the second added every second.
15. The distances traversed during equal intervals of time, by a body falling from rest, stand to one another in the same ratio as the odd numbers beginning with unity.

The social studies are concerned primarily with the origin of these principles, their significance in the development of thought, and their contribution to social progress. But in the social studies materials they are isolated facts taken from carefully formulated sequences of facts. It is the province of mathematics to make clear the relationship of these principles in their proper sequence and to create an awareness of their place in individual and social development. An assurance that pupils are able to appreciate a mathematical principle in other than mathematical literature requires more than the knowledge that they are familiar with the word concepts involved. Such an assurance comes from confidence in carefully planned and thorough mathematical instruction. Undoubtedly, the social studies will use mathematics to an even greater advantage when there is more recognition of the interrelationship between mathematics and the social studies, and when teachers of mathematics teach more *about* mathematics in the teaching of mathematics.

A second significant use of mathematical terms in the social studies materials occurs in the discussions of the nature of mathematics. This usage tends to strengthen the conviction that training in mathematics that is desirable as preparation for the study of the social sciences must emphasize its nature.

"When we think of mathematics, we have in our mind a science devoted to the exploration of number, quantity, geometry, and in modern times also including investigations into yet more abstract concepts of order, and into analogous types of purely logical relations. . . .

". . . the applicability of mathematics to an understanding of our world is removed for many minds because they very inadequately appreciate the great abstractness of mathematics. They do not realize that the geometrician, in proving his propositions, claims to be discoursing of nothing that ever was on land or sea; that the arithmetician, when he states that $2+2=4$, asserts a certain relation between perfectly abstract concepts which would still hold if two things could never be added to two other things without creating a fifth thing, or without combining to form a single thing. No matter how things may combine, $2+2=4$ because of the meaning which the mathematician has given 2, +, =, and 4. . . .

"Every mathematical system rests upon certain postulates and assumptions which are simply taken for granted without proof, and from which the rest of the system follows with strict logical rigor." [2]

The power of mathematics, too, is described in the materials of the social studies.

"Descarte's invention of analytic geometry offers a stunning example of the simplicity of many great discoveries. It had long been a matter of almost common knowledge that the position of a point on a plane is determined when its relation to two known points has been established. . . . Descartes did nothing more than apply this fact to the description of the path of a point, that path being referred in a fixed manner to two intersecting lines (co-ordinates), but as a result it became possible to give algebraic equivalents for geometrical figures, and a whole new field of mathemat-

[2] Storck, John, *Man and Civilization,* p. 394. 1927.

ical endeavor was opened up. The calculus follows from analytical geometry as a natural corollary, although the fact that the simple step forward was not taken for over a generation indicates the presence of forces working against the new development."[3]

The discussions in the social studies are concerned not only with the nature of mathematics as a basis for describing its place in the evolution of society, but also with its place in the development of thought.

"A reflective mind observing the identity of the angle in the many positions in which the square could be drawn could, one would think, in the end not resist the conclusion that the 'angle in a semicircle is a right angle.'

"When man came to the circle, he had already perceived in a vague, unanalyzed way the common quality of perfect roundings. We cannot believe that any animal has this perception, and the lowest savage has not expressed it.

"The next step comes when the drawing of the circle elicits the latent knowledge of its most obvious property, that the circumference is the locus of all the points touched by the end of a string or stick revolving round the center. So far prescientific man has gone; the first Theorem of Thales is but another step in the analysis.

"The perception itself of the right angle in the semicircle does not appear much more difficult than that of the equality of radii; its wider scope arises from its formulation in exact and general terms, and from the circumstances that the observation brings together two distinct classes of figures, triangles and circles, and sets up a universal relation between them."[4]

The accounts of the growth in the knowledge of natural events requires the use of numerous mathematical concepts.

Copernicus "postulated that the universe was spherical, and retained the old Ptolemaic sphere of the fixed stars; he also postulated that the motions of the celestial bodies are uniform circular motions or are compounded of such motions—Ptolemy's theory of epicycles . . . The startling thought contained in his work was . . . only reason operating by mathematical calculation could be trusted. . . .

[3] *Ibid.*, p. 278. [4] Marvin, F. S., *The Living Past*, p. 60. 1931.

" . . . the accurate records of Tycho had revealed the impossibility of calculating the orbits of the planets by Copernicus' combination of circular movements; so he tried combination after combination of new circles, calculating the results of each and comparing them with the records. One geometrical scheme coincided with an error of but 8′, and he was tempted to rest here. . . .

"He tried next, not a circle, but an oval, attempting to fit the data from Mars into an egg-shaped orbit. Then he turned to the simplest known oval curve, the ellipse, and the success of his attempt confirmed his faith in 'the simplicity and ordered regularity of Nature.' He had discovered at last the true shape of the planet's orbit, an ellipse with the sun in one focus." [5]

The consideration of refinements in measurement and the measurement of individual differences and abilities include mathematical terms.

"Now it is possible without trouble to bore a cylinder 10 feet in diameter with a variation in diameter of less than one-hundredth of an inch, and much greater accuracy can be obtained if necessary." [6]

"By Neilson's invention, when the blast was heated to 300° F., 5 tons of fuel were needed to produce one ton of product, as against 8 before; while but 2¼ tons were required when the blast was heated to 600° F." [7]

"Individuals differ from each other because the determiners (environmental and heredity) of any given trait are always both numerous and small, so that the probability of any particular number of them being present in a given instance is in accord with the laws of chance. Suppose, for example, that a man's height were determined by the co-operation of twenty elements, each of equal importance, and that the probability of any single element being present were just as great as the probability of its being absent. In that event, it is clear that on the average ten of the elements would be present and ten absent, while the probability of any other assigned number being present in a given instance would be calculated according to the law which states the frequency with which

[5] Randall, John H., *The Making of the Modern Mind,* p. 229. 1926.

[6] Storck, *op. cit.,* p. 211.

[7] *Ibid.,* p. 209.

a given number of heads will appear when twenty coins are tossed into the air. . . .

"The following features, then, will characterize practically all instances where individual differences are found to occur:

"(1) These differences will nearly always cluster around a type or mode, and large variations from the mode will be less common than small, becoming rarer and rarer as the size of the variations increases. . . . Statisticians have devised a number of measures for summarizing the known numerical facts about such distributions. Thus one may report: the total number of cases measured, with the measurement for each; the range, or numerical differences between the lowest record and the highest for the group; the average or central tendency of the group (calculated in several different ways); the degree of scatter or deviation from the average; and sometimes also the skewness, or degree to which the data depart from such chance distributions as one would obtain by tossing pennies (this distribution being taken as typical or normal).

"(2) Variations are usually continuous; . . .

"(3) Two differently selected groups will seldom contain precisely the same distribution of individual measurements, though it usually happens that the measurements for the two groups will overlap, the range of measurements within either group being greater than the difference between the averages of the groups. . . .

"(4) If two or more qualities exhibited by the different members of the group are measured, it will nearly always be found that the qualities in question to some degree depend on each other. As the one varies in amount, the other varies also.

"Statisticians have devised a measure (called the coefficient of correlation—denoted by r) of the degree to which given qualities coexist in the different members of a group. This measure varies in value between $+1$ and -1; it is $+1$ when the relationship is complete and positive, so that for every occurrence of a specific amount of the first quality in a given individual there is a corresponding occurrence of the second quality in that individual; it is -1 when for every instance of the first quality there is a corresponding opposite instance of the second; and it is 0 when there is no correspondence whatever between the two qualities." [8]

Two further quotations show uses of mathematics involving variation and units of measure in the content of the social studies materials.

"But, if this medium (the ether) exists, nothing definite is known about it except that it transmits radiation which . . . consists of transverse waves. If it is an elastic medium, the velocity of the transverse waves in it is equal to $\sqrt{E/D}$, where E is its transverse elasticity or rigidity and D its density; and since the velocity of light is enormously great, either E is very great or D is very small. If E is very great, the medium is practically rigid, and it is difficult to see how the planets and comets can move freely through it. If D is very small, the ether is much less dense than any gas, and it is not easy to understand how it can have rigidity at all." [9]

"Man has developed entirely artificial units, such as the hour, the minute, the second, and the week, upon the basis of such natural time spans as the day, the month, and the year.

"One of the chief functions of a precise time notation is its usefulness for the recording of scientific observations, many of which depend upon our ability to make very small (and in some cases very large) temporal discriminations. In connection with scientific research, time and space have been minutely subdivided. Thus the Hipp chronoscope, which records thousandths of a second, is used in psychology . . . , while the interferometer, a far more precise instrument, is employed to measure possible minute differences in the speed of light waves. . . . On the other hand, units have been developed for the measurement of geological periods lasting for millions of years. . . . Space has been similarly divided, from such minute measures as the Angstrom unit (which equals 10^{-10} meters, or one millionth of .003937 inch), up to such distances as the light year, or the distance traversed by a ray of light in a year when traveling at a speed of 186,000 miles a second. The progress of scientific discovery would be severely hindered if

[8] *Ibid.*, p. 154.
[9] *Ibid.*, p. 408.

these measures, as well as others for volume, weight, density, pressure, energy, etc., were lacking." [10]

The mathematical terms involved in the social studies in the statements of certain mathematical principles derived in the study of geometry and in the study of physics, in the exposition relating to the nature of mathematics and its importance as a method of thinking, in several descriptions of the place of mathematics in interpreting nature and in the development of science and scientific procedure, and in connection with simple and more precise practical measurements and the corresponding measuring instruments in common use and in the measurement of the more abstract qualities of individuality and personality, continue to give evidence of the multiplicity of interrelations between mathematics and the social studies materials. The separation of the mathematical terms from the social studies content can serve to indicate to a certain limited extent the mathematical knowledge required to comprehend the social studies materials. But it must be emphasized again that the mere understanding of the meanings of the terms involved is no assurance of attaining that fullness of appreciation and interest "necessary to the general understanding and management of an integrated society in its historic and world setting" [11] as one of the primary functions of instruction in the social studies. It is necessary to consider for the maximum amount of carry-over the setting in which the terms are learned, and it is necessary to consider, as a part of the requisite mathematical training, the amount of marginal mathematics essential in acquiring the full meaning of the terms.

The mathematics that has been considered in this chapter has been limited to the mathematics found in an examination of the social studies materials suggested in a single source. These materials are presumed to include textbooks differing from each other sufficiently in subject matter and in organization to afford a fair criterion for judging the type of mathematics that pupils who pursue the elementary prescribed courses in the social studies at the college level may be expected to encounter.

[10] *Ibid.*, p. 185.

[11] *Report of the Commission on the Social Studies, Conclusions and Recommendations of the Commission,* p. 40, 1934.

Additional materials would undoubtedly employ mathematical concepts and processes not already listed and emphasize the uses of certain aspects of mathematics that were not stressed in the materials examined.[12] But the mathematical training necessary to care for this variation, to the extent that it has been recognized in this study, does not appear to be essentially different from that found necessary to understand the mathematics in the textbooks examined. The growth in the mathematical needs of the social studies is in evidence more in the works of Fisher, Edgeworth, Bowley, Jevons, Mitchell, and others beyond the scope of this study than in the elementary prescribed courses.

Another contact between mathematics and the social studies is evidenced by the presence of the names of many celebrated mathematicians. Among the mathematicians whose contributions to mathematics or to social progress are discussed in the social studies are the names of Albertus Magnus, Anaxagoras, Apollonius, Archimedes, Aristarchus, Aristotle, Bernoulli, Bowditch, Brahe, Briggs, Cardan, Cauchy, Cavalieri, Clairaut, Condorcet, Copernicus, D'Alembert, Democritus, Descartes, Diophantus, Einstein, Eratosthenes, Euclid, Eudoxus, Euler, Fermat, Galileo, Green, Gregory, Herodotus, Hipparchus, Hippocrates, Hooke, Huyghens, Jordan of Saxony, Kelvin, Kepler, Lagrange, Laplace, Leibnitz, Leonardo Pisano, Maupertuis, Maxwell, Napier, Newton, Pascal, Peuerbach, Plato, Poincaré, Ptolemy, Pythagoras, Regiomontanus, Roemer, Stokes, Thales, Torcelli, and Vieta.

SUMMARY

The tabulation of the mathematical concepts that occur in the materials of instruction for the prescribed introductory college course in the social studies shows that the concepts of number and measurement are those commonly found in textbooks of the elementary arithmetic or junior high school mathematics, or concepts that may be expected to be easily comprehended with mental maturity and extensive practice with those that do occur in the textbooks. Occasionally the symbolism of algebra occurs.

[12] See, for example,

Ogburn, W. F. and Goldenweiser, A., *The Social Sciences and Their Interrelations,* 1927.

Smith, Preserved, *A History of Modern Culture, Vols. I and II.* 1930.

Mathematical terms occur more frequently and with greater diversity in these materials than in the materials examined for the secondary school. A comparison of these terms with the defined terms in the arithmetic and junior high school mathematics textbooks examined shows that the texts do not provide for the acquiring of the meanings of a number of terms associated with (1) geometry and the geometrical method, (2) statistics, (3) measuring instruments, (4) advanced mathematics, and (5) the names of the branches of higher mathematics. There is no implication that these terms should be provided for in the teaching of mathematics below the tenth year. It is more probable that mathematical training considerably beyond the commonly regarded minimum must be considered as desirable preparation for the study of even the elementary social studies materials at the college level for adequate comprehension of the materials. Such training would assist in correcting the situation that "verbalism, the use of words without understanding, has cursed the school all down through the centuries and, as investigations by the commission prove, continues to corrupt social science instruction in American schools today."[13]

[13] *Report of the Commission on the Social Studies, Conclusions and Recommendations of the Commission,* pp. 78-79, 1934.

CHAPTER V

MATHEMATICS IN ELEMENTARY ECONOMICS

RELATION OF MATHEMATICS AND ECONOMICS

"ECONOMICS deals with the production, exchange, possession, consumption, and use of material goods and immaterial services. The whole subject of wealth and welfare has two aspects, one subjective, moral or psychological, the other objective or material. From the one we may consider the attainment by economic action of an abstract good, or hedonistically the pleasure or satisfaction derived from the possession or use of things, or the desire to obtain goods; none of which terms are arithmetically measurable. From the other we may have in view material goods and actual services which can be measured by quantity or by money value. At first sight it might appear that mathematical reasoning was confined to the objective aspect, but this is not the case. If we cannot measure, it is true that we cannot apply the arithmetical processes of addition and multiplication and their converses; but we may be able to detect equality and inequality, relationship, continuity, variation, and other properties which lead to algebraic expressions." [1]

SELECTION OF THE MATERIALS OF INSTRUCTION IN ELEMENTARY ECONOMICS

The courses in economics offered in the senior high schools and the first courses in economics offered in colleges and universities are considered in this study as courses in elementary economics. The findings in this chapter are based on an examination of five textbooks in economics that were found in common use by three recent studies.[2] The findings represent, as in Chapter IV, the view-

[1] Bowley, A. L., *The Mathematical Groundwork of Economics*, p. 1. 1924.

[2] Lomax, P. S. and Tonne, H. A., *Problems of Teaching Economics*. 1932.

Rugg, E. U. and Dearborn, N. H., *Social Studies in Teachers Colleges and Normal Schools*. 1928.

Shields, H. G., "School and College Courses in Economics," *Journal of Educational Sociology*, November, 1930, pp. 155-157.

point, the use of mathematics, and the mathematical concepts in the sources from which they are taken. Because of some recent changes, the results may differ somewhat from present thought or present numerical values, but from the standpoint of the mathematical training involved, these differences cannot be regarded as significant. The complete list of books examined is given in Appendix V.

RESULTS OF THE ANALYSIS OF THE MATERIALS

Much of the mathematics involved in the exposition throughout the study of economics in all its divisions of production, consumption, exchange, distribution, and public finance is not different from that listed in the previous chapters for the more general social studies courses at the junior high school and at the college levels. A study of the exact extent of the occurrence of the concepts in elementary economics similar to those discussed in Chapters II and IV was confined to a single high school textbook. The frequency and nature of the numbers that were found are shown in Table XXVI. The frequency and nature of the units of measure in use are given in Table XXVII. The monetary units and the units of time are the most frequently used units. Of some, but less frequent, use are the units of weight, capacity, and area.

TABLE XXVI

FREQUENCY AND NATURE OF NUMBERS IN THE ELEMENTARY ECONOMICS TEXTBOOK EXAMINED

Type and Range of Numbers	Frequency
Ordinal numbers in figures 1 to 8,476	1,997
Ordinal numbers in words first to hundred-and-first	296
Roman numerals I to XXXI	96
Dates	748
Cardinal numbers in figures 0 to 320,804,000,000	1,170
Cardinal numbers in words zero to trillions	1,176
Mixed numbers 1 1/5 to seven and one half billions	48
Fractions three one-hundredths to nine tenths	100
Decimals 0.10 to 134.39	163
Total	5,794

Table XXVIII contains three lists of terms that were found in the analysis and separated from the general mathematical vocabulary for consideration. The first list contains seventeen terms that are often used in connection with the results of mathematical computation, but by definition are not necessarily confined to the results of computation. Selling price may involve no computation whatever, or it may represent a term in a mathematical formula. Interest is money paid for the use of money and as such may be obtained through the use of a complex formula, or, again, with no

TABLE XXVII

FREQUENCY AND NATURE OF UNITS OF MEASURE IN THE ELEMENTARY ECONOMICS TEXTBOOK EXAMINED

Unit	Frequency	Unit	Frequency
Acre	11	Half-eagle	1
Barrel	1	Hour	21
Board foot	2	Long ton	7
British thermal	1	Mile	7
Bushel	49	Minute	1
Calorie	1	Month	15
Cent	72	One-cent	1
Century	29	Ounce	17
Cwt.	2	Pound (sterling)	6
Day	73	Pound (weight)	12
Decade	4	Quarter	2
Degree (arc)	1	Quarter-dollar	1
Dollar	415	Quarter-eagle	2
Double-eagle	1	Score	1
Dozen	7	Shilling	1
Dime	2	Square mile	1
Eagle	1	Ton	9
Five-cent	1	Week	12
Franc	1	Yard	2
Grain	9	Year	196
Half-dollar	3	Total	1,001

computation whatever. Similarly, the second list contains terms associated with mathematical quantities, but less clearly the outcomes of numerical calculations. The third list contains terms that appear frequently in the explanations of mathematical processes and are the names of commercial or business instruments. Many of these concepts are defined and clarified in the content of the social studies and in the content of the arithmetic. In practice the terms are introduced somewhat earlier in arithmetic than in the social studies. The social studies should profit from this earlier introduction, if the relationship is emphasized in arithmetic.

TABLE XXVIII

VOCABULARY ASSOCIATED WITH MATHEMATICS IN THE ELEMENTARY ECONOMICS TEXTBOOK EXAMINED

Term	Frequency	Term	Frequency	Term	Frequency
Amount	74	Debt	21	Account	18
Annuity	1	Deposit	15	Bill of exchange	12
Balance	28	Duty	3	Bill of lading	8
Capitalization	1	Fee	2	Bond	28
Cost	53	Income	87	Budget	3
Discount	7	Insurance	35	Check	16
Dividend	13	Liability	9	Common stock	17
Interest	93	Loan	15	Credit	32
Loss	26	Price	358	Draft	10
Margin	2	Rent	80	Installment	5
Net profit	4	Resource	6	Maturity	2
Par	10	Salary	12	Money	200
Premium	1	Surplus	41	Mortgage	5
Principal	6	Tariff	49	Note	10
Profit	98	Tax	119	Preferred stock	8
Rediscount	2	Wages	91	Promissory note	5
Selling price	3			Share	9
				Statement	3
				Stock	27
				Stock certificate	1
				Trade acceptance	5
Total	422		943		424

The mathematical terms that appeared with the largest frequencies in the earlier years of the study of the social sciences continue to occur in economics with correspondingly large frequencies and similar connections. The concepts appearing with large frequencies are per cent, rate, and average. The newer concepts requiring an understanding of their meanings, include those associated with index numbers, with function graphs, and with variation. The mathematical terms are listed in Table XXIX. The remaining tables contain the results of an analysis of the five textbooks.

Economic life consists of relationships and the functioning of relationships. The type of thinking involved in the study of economic life "lacks the exactness of mathematical demonstration; but the average person has practical need, also, for another method of thought or reasoning. In life we deal not with certainties but with tendencies and probabilities. In economics we do the same."[3]

[3] Fay, Charles R., *Elements of Economics*, p. 3. 1932.

TABLE XXIX

General Mathematical Vocabulary in the Elementary Economics Textbook Examined

Term	Frequency	Term	Frequency
Accelerate	2	Increment	2
Acceleration	2	Indefinite number	1
Add	11	Indeterminate number	1
Addition	8	Index number	8
Aggregate	7	Infinitely	2
Angle	3	Inversely	2
Apex	1	Limit	3
Apportion	6	Line	3
Approximate(ly)	14	Mathematically	1
Approximation	2	Mathematics	2
Area	7	Measure	25
Average	59	Measurement	1
Base	1	Minus	9
Calculate	1	Multiplication	1
Calculation	4	Multiply	3
Coincide	1	Parallel	1
Common denominator	1	Per cent	251
Compounded	2	Percentage	3
Compute	1	Plus	18
Computation	1	Point	7
Constant	3	Proportion	6
Converge	1	Proportionate	3
Cube	1	Pyramid	1
Curve	10	Rate	125
Cylinder	1	Ratio	10
Date	2	Rectangle	10
Decrement	1	Relative numbers	1
Difference	7	Residual	1
Disproportionate	1	Slope	1
Divide	15	Statistical	1
Divisible	1	Square	2
Division	38	Subdivide	2
Equal	48	Subtract	1
Estimate	14	Subtraction	1
Figure	4	Sum	16
Fixed	5	Times	6
Fractional	1	Unit	88
Graphically	4	Vary	7
Height	3	Vertical	2
Horizontal	1	Volume	16
Total			939

The problem of the mathematician in the field of economics is the degree of exactness to which the relationships in economics are capable of being put into mathematical language and of being represented by mathematical methods.

The problem of measurement in economics is of much importance. "Economic goods and wealth are quantitative concepts; they must be measured." "Exact measurement is unnecessary for many important conclusions." "Whether we shall ever be able to measure the variation of marginal utility between typical families from time to time or place to place is doubtful. But there can be no doubt of the deep scientific and practical need for such measurements; and no valid objection arises from the fact that the total utilities are immeasurable and incommensurable. At the origin wants and want satisfactions are incommensurable. At the margin want satisfactions are commensurable." "Even those laws of uniformities which the science prizes as the finest product of its research are but statements of probabilities."[4] These quotations evidence the fact that the economists appreciate the importance and need of measurement in their science and are cognizant of the difficulties involved in measurement where measurement is found desirable. They do not hesitate to make use of measurement to the degree of accuracy possible with the difficulties encountered, but they are cautious in interpreting the results obtained. They are careful to point out, where necessary, that statistical comparisons represent "broad tendencies" and not "accurate measurements."

Measurement results in the use of decimals. The use of decimals in the economics materials is considerably more extensive than in the more general social studies materials. The frequencies and the nature of the decimals that appeared in each of the five textbooks examined and the connections with which they appeared are shown in Table XXX. A, B, and C are used to denote the high school textbooks examined, and D and E the college textbooks. About 15 per cent of the 5,438 decimals are pure decimals, and approximately 96 per cent are limited to one, two, or three decimal places. Ratios of exchange were expressed to more than three decimal places, and occasionally per cents to show significant figures.

Equations and formulas are developed as a result of measure-

[4] Ely, Richard T. and Others, *Outlines of Economics*, Chap. VI. 1931.

TABLE XXX

FREQUENCY OF DECIMALS WITH THEIR CONNECTIONS IN THE ELEMENTARY ECONOMICS TEXTBOOKS EXAMINED

Type of Decimal	Frequency: Textbook A	B	C	D	E	Total	Connections
.1*	1	22	1	37	21	82	acre, average, billion, bushel, cent, dollar, grain, index number, million, month, per cent, persons per thousand, quotient, thousand, times, year.
1.1	16	103	19	170	210	518	
11.1	21	436	61	307	454	1,279	
111.1	1	695	38	110	176	1,020	
1111.1		9	28	20	4	61	
11111.1		14		1		15	
111111.1		7				7	
1111111.1		6				6	
Subtotal	39	1,292	147	645	865	2,988	
.01		26	2	51	27	106	average, cent, dollar, grain, hour, per cent, quotient, thousand.
.11	31	73	26	97	77	304	
1.11	40	181	61	197	121	600	
11.11	36	184	32	123	100	475	
111.11	4	171	1	28	13	217	
1111.11		155		1		156	
11111.11		77	1	1		79	
111111.11		44				44	
1111111.11		20				20	
11111111.11		19				19	
111111111.11		11				11	
1111111111.11		2				2	
Subtotal	111	963	123	498	338	2,033	
.001		1		4		5	average, billion, cent, dollar, per cent, quotient.
.011		4		4	1	9	
.111		38		13	56	107	
1.111	13	8	7	30	13	71	
11.111				20	5	25	
Subtotal	13	51	7	71	75	217	
.0001		13		2		15	cents, dollar, grain, per cent, quotient.
.0011		9		3		12	
.0111		42	5			47	
.1111		85				85	
1.1111		8	26			34	
11.1111			4			4	
111.1111		1	1			2	
Subtotal		158	36	5		199	
.00111		3				3	dollar
.01111		7				7	
.11111		10				10	
Subtotal		20				20	
1.111111		1				1	dollar
Total	163	2,465	313	1,219	1,278	5,438	

* The (1) is used for significant figure.

ment, but involve more than measurement. The equation "has the great virtue of emphasizing the interdependence of the factors included. Change any factor in the equation and the other factors must change unless they happen to counteract each other."[5] The twenty equations and formulas encountered are presented in Table XXXI with their connections in the economics materials. The equation of exchange will serve to illustrate the use of the equation in economics.

"The value of money tends to vary inversely with the quantity of currency and its velocity of circulation and directly with the volume of trade."

The relations in the form of an equation result in the algebraic equation

$$MV + M'V' = PT$$

where M, V, M', V', P, and T have the meanings explained in Table XXXI.

"The quantity of money multiplied by its velocity of circulation is evidently the total amount of money paid in making exchanges during the year. Likewise the quantity of deposits multiplied by their velocity of circulation gives the total amounts of payments by transfer of deposits. The sum of these two products is the total amount of payments, in all kinds of currency. This magnitude and the total amount of currency payments are necessarily the same, and the 'equation of exchange,' as it is called, is the algebraic expression of this relation.

"It is P which measures the general level of prices and (inversely) the value of money, and the effect upon P of any of the other magnitudes is clearly shown by the equation. If we direct our attention to the two most important factors, M and T, it is clear that changes in T tend to cause opposite changes in P; which is the algebraic expression of the general conclusion"[6] stated above.

The understanding of the formulas and equations requires an understanding of the symbolism involved. In each case the literal algebraic symbols were defined. The multiplication of two literal

[5] *Ibid.*, p. 296.

[6] Fairchild, F. P., Furniss, E. L., and Buck, N. S., *Elementary Economics*, Vol. I, p. 512. 1932.

TABLE XXXI

EQUATIONS AND FORMULAS, INDEX NUMBERS, POSITIVE AND NEGATIVE NUMBERS, STATEMENTS, AND DIRECT AND INVERSE VARIATION IN THE ELEMENTARY ECONOMICS TEXTBOOKS EXAMINED

Descriptions and Connections	Textbook A	B	C	D	E
Equations and Formulas					
Of exchange					
$M \times V = p \times Q$		x			
$M \times V + M' \times V' = p \times Q$		x			
$M \times V = P \times T$				x	
$M \times V + M' \times V' = P \times T$		x		x	x
$\frac{V}{T} = K$				x	
$M = \frac{P}{K}$				x	
M —the amount of money					
V —the velocity of circulation of money					
Q —the quantities of goods					
P —the scale of prices					
T —the volume of trade					
p —the prices of goods					
M'—the amount of "credit currency"					
V'—the velocity of circulation of "credit currency"					
K —a constant					
Of price level					
Price level = $\frac{\text{supply of money}}{\text{demand for money}}$			x		
Price level = $\frac{\text{amount of money x rate of turnover}}{\text{demand for money}}$			x		
$P = \frac{M \cdot R}{V}$			x		
P —the price level					
M —the amount of money					
R —the rate of turnover					
V —the demand for money					
Of compound interest					
$A_n = P(1 + r)^n$					x
A —the future value					
P —the present value or principal					
r —the rate of interest					
n —the number of years					
Of present worth					
$P = \frac{i}{r}$					x
P —the present worth					
i —the periodical payment					
r —the rate of interest					

TABLE XXXI (Continued)

Descriptions and Connections	Textbook A	B	C	D	E
Of production					
$P = L^{3/4} C^{1/4}$				x	
$P = L^{.6} C^{.4}$				x	
P —product					
L —labor					
C —capital					
Of railroad costs					
Total rate $= t + rm$				x	
t —terminal charge					
r —charge for each mile of line haul					
m —number of miles					
Of the rectangular hyperbola					
$xy = a$					x
a —a constant					
x —a variable					
y —a variable					
Of index numbers					
The index number is $\frac{\Sigma(q_0 + q_1)p_1}{\Sigma(q_0 + q_1)p_0}$				x	
The ideal index number is $\sqrt{\frac{\Sigma p_1 q_0}{\Sigma p_0 q_0} \times \frac{\Sigma p_1 q_1}{\Sigma p_0 q_1}}$				x	
$p_0, p_0^1, p_0^2 — p_0^n$ — the prices of the various commodities in the base year					
$p_1, p_1^1, p_1^2 — p_1^n$ — the corresponding prices in year to be compared with the base year					
$q_0, q_0^1, q_0^2 — q_0^n$ — the respective quantities in which the commodities are purchased in the base year					
$q_1, q_1^1, q_1^2 — q_1^n$ — the corresponding quantities in year of comparison					
Of price					
Cost + profit = selling price			x		
Of capital					
Assets — liabilities = capital			x		
Of geometric average					
The nth root of the product of the relative prices of n commodities		x		x	
Of profits					
Differential or pure profits = (Gross Profits) — (entrepreneur's interest, rent, and, possibly, wages of management + replacement fund, surplus fund + monopoly gains, "conjunctural gains")		x			

TABLE XXXI (Concluded)

Descriptions and Connections	Textbook A	B	C	D	E
Index Numbers					
Cost of living		x			
Employment			x		
Expenditures				x	
Income				x	
Marketing		x			x
Prices	x	x	x	x	x
Production	x			x	x
Stocks					x
Positive and Negative Numbers					
On the axes of graphs			x	x	
To denote direction of change in balances and in net revenue		x		x	x
To denote direction of rates of change				x	x
Statements					
Bank	x	x	x	x	x
Corporation					x
Individual			x		
Partnership			x		
Profit and loss					x
Trade	x				
Direct and inverse variation					
Between cost and business		x			
Between demand and price	x	x	x	x	x
Between demand and supply	x	x	x	x	x
Between income and expenditures		x			
Between amount of interest and rate of interest	x				
Between price and purchasing power		x		x	
Between purchasing power and demand		x			
Between rent and price	x				
Between supply and price	x	x	x	x	x
Between utility and demand		x			
Between utility and supply					x
Between value and price	x		x		x
Between value and supply	x		x		
In the Law of Diminishing Returns	x	x	x	x	x
In the Law of Proportionality		x	x	x	x

quantities occurred, indicated by the usual times sign, the dot, or with the absence of a sign of operation. Other processes involved division, addition, and subtraction; the use of the radical; parentheses; and exponents. The development, manipulation, substitution, and simplification of the formula occurred.

The index number is another significant means of measurement in common use in economics. Index numbers have the advantage of permitting the measurement of average or of general changes in things which are measured in different units. Not all index numbers are equally reliable. "By means of index numbers, changes in the purchasing power of money to well-defined classes of consumers may be measured with an exactness which approaches that attained by ordinary physical measurements of weight or length. But as a measure of change in the general price level, the index number of prices has essential limitations. Grave technical difficulties are involved in computing an index number." "It is, moreover, exceedingly hard to handle such a figure intelligently after it has been computed."[7] The meaning of and the method for calculating index numbers occurred in all the economics textbooks that were examined. The uses of index numbers are shown as they occurred in economics in Table XXXI. Methods suggested for computing index numbers are shown in Table XXXII.

Measurement involves positive and negative numbers. Their use in economics is not frequent. It was confined to measurement along the vertical axis of graphs and in indicating the nature of balances and rates of exchange. The fundamental notion of positive and negative numbers occurs in other situations. For example, "a $1000 bond is positive wealth to the creditor but negative wealth to the debtor, and the two cancel out substantially when national or social wealth is estimated."[8]

The concepts of variation and dependence are used widely in the discussion of economic principles. The understanding of variable and constant, and direct and inverse is necessary. The law of the proportion of factors in economics illustrates the use of the terms. "When the law of the proportion of factors is stated, one factor is usually taken to be constant and another variable; for example, a fixed area of land and a variable amount of labor. This means that for the purposes of short-run problems, at any rate, land represents a constant cost and labor a variable one. As labor per acre increases, the crop per acre naturally passes through five stages. It may increase (1) faster than labor, (2) just as fast, (3) slower, (4) not at all, or (5) it may actually decrease. The first

[7] Ely and Others, *op. cit.*, p. 234.
[8] *Ibid.*, p. 95.

case represents increasing returns to labor; the second represents constant returns, the third diminishing returns, the fourth zero returns, and the fifth less than zero returns."[9] It is explained that the terms inversely and directly, as used in economics, do not imply "that the variation must necessarily be proportional. If *a* increases when *b* increases or decreases when *b* decreases, we say that *a* varies directly with *b*, even though the change in *a* may not be exactly proportional to the change in *b*; and correspondingly of inverse variation."[10] The uses of variation in economics are listed in Table XXXI. The law of demand is an example of variation. "In a given market at a given time, the quantity of any goods which buyers would choose to take, when not a fixed quantity, varies inversely with the price."[11] The quantity theory of money provides a similar example in different terms. "An increase or decrease of the quantity of money, the volume of trade being constant, must be accompanied by a proportionate general increase or decrease in prices."[12] Another example occurs in the law of supply. "The supply is always relative to a particular price and, broadly speaking, varies directly with price, although not proportionately."[13]

The mathematical processes and computations that were carried out in the economics materials examined are shown in Table XXXII. The table shows only those processes in which the computations were actually illustrated in the exposition. The computations in the table are limited to one illustration for each distinct process, since the detail in which the computations were carried out represented the only significant differences in the different books. The number of different books in which the computations occurred is indicated on the right side of the table.

The mathematical activities required in economics are classified in Table XXXIII. The activities were confined to three of the textbooks examined, one of the textbooks containing no exercises of any kind and the other no exercises involving mathematical activities. Although the majority of the activities have to do with

[9] *Ibid.*, p. 121.
[10] Fairchild, Furniss, and Buck, *op. cit.*, p. 260.
[11] *Ibid.*, p. 261.
[12] Ely, and Others, *op. cit.*, p. 298.
[13] Faubel, Arthur L., *Principles of Economics*, p. 82. 1932.

TABLE XXXII

MATHEMATICAL COMPUTATIONS APPLIED IN THE ELEMENTARY ECONOMICS TEXTBOOKS EXAMINED

Computation and Application	Textbook A	B	C	D	E
Averages:					
Simple arithmetic average	x	x	x	x	x
$\frac{8 + 13 + 16 + 18}{4} = 13\frac{3}{4}$					
Weighted average		x		x	x

	Units sold	Price per Unit	Gross values
Iron	10,000	$40.00	$400,000
Wheat	50,000	2.00	100,000
Cotton	100,000	0.15	15,000
Wool	200,000	0.12	24,000
Total	360,000		$539,000

Weighted average for the year is $539,000 ÷ 360,000 = 1.496

Computation and Application	A	B	C	D	E
Geometric average				x	
The geometric mean is $\sqrt{200 \times 50}$, or 100					
Balances	x	x	x	x	x
The balance was $232,000,000 ($962,000,000 minus $730,000,000)					
Book Value					x
The book value is now $156.256 (i.e., $1,562,560 divided by 10,000)					
Capital			x		
Assets − Liabilities = Capital: $100,000 − $50,000 = $50,000					
Capitalization		x	x	x	x
If the rent of a parcel of land is $600 and the prevailing rate of interest is 6%, 1% will be $100, and 100% will be $10,000; the value of the land will be $10,000.					
Comparative Cost			x		
England's cost of cloth is $6; of the rails, $30. In England, cloth costs 1/5 of what rails do, $6/$30; rails cost five times the cost of cloth; $30/$6. England's comparative cost of cloth is 1/5; her comparative cost of rails is 5.					

TABLE XXXII (Continued)

Computation and Application	Textbook A	B	C	D	E
Denominate Numbers			x		
15 yards of cloth + 2 tons of steel 10 yards of cloth + 1 ton of steel ——————————— 25 yards of cloth + 3 tons of steel					
Discount			x	x	x
The bank determines what sum put at interest for 60 days at 6 per cent will amount to $50,000; namely $50,000 ÷ 1.01, – which is $49,504.95. The $495.05 represents the true discount.					
Dividends	x				x
10,000 shares 6% preferred stock, par value $100 a share 10,000 shares common stock, par value $100 a share Net earnings $150,000 Requirements for preferred stock dividends 60,000 ——————— Available for common stock dividends $ 90,000 Dividend payment per common share $9					
Exchange			x	x	x
Par of exchange The pound sterling contains 113.0015 grains of pure gold The dollar contains 23.22 grains of pure gold The par of exchange is found by dividing 113.0015 by 23.22, which is 4.8665 Equation of exchange 5,000,000 x 20 times a year = 200,000,000 loaves x $.10 a loaf + 10,000,000 tons x $5.00 a ton + 30,000,000 yards x $1.00 a yard					
Index Numbers	x	x	x	x	x
Suppose the weighted averages for successive years are 1.496, 2, 2.1, 1.8, and 1.3 (See weighted average.) If 1.496 is 100, then 2 is 2 ÷ 1.496 x 100 = 133 2.1 is 2.1 ÷ 1.496 x 100 = 140 1.8 is 1.8 ÷ 1.496 x 100 = 120 1.3 is 1.3 ÷ 1.496 x 100 = 86 The index numbers are 100, 133, 140, 120, and 86.					
Interest	x	x	x	x	x
Simple interest The interest (on $1,000) for 365 days at 4% is $40; for one day it is .109; for 31 days it is $3.38.					

TABLE XXXII (Continued)

Computation and Application	Textbook A	B	C	D	E
Compound interest When the interest rate is five per cent, \$100 invested now will be worth \$105; i.e., \$100 x 1.05, one year from now; \$110.25; i.e., \$100 x $(1.05)^2$, two years from now; \$115.76; i.e., \$100 x $(1.05)^3$ three years from now; and so on.					
Percentages				x	
$\frac{1.17}{1.27} = .92$, or 92%					
Present Worth				x	
\$100 due one year hence is worth now \$95.23; i.e., \$100 ÷ 1.05, assuming an interest rate of five per cent; if due two years hence, it is worth now \$90.70; i.e., \$100 ÷ $(1.05)^2$; and so on.					
Progressions		x	x		
Arithmetic progression 2, 4, 6, 8, 10, 12, 14, etc. Geometric progression 2, 4, 8, 16, 32, 64, 128, etc.					
Ratios	x	x	x	x	x
Bimetallism The legal ratio was $\frac{371.25}{23.2}$ or $\frac{16.002+}{1}$ Exchange The ratio is $\frac{113.00106\ 2/3}{23.22}$ or 4.8665+ Relative prices Wheat : eggs : : 6 : 1 Reserve The reserve ratio is now $\frac{292,410}{1,465,083}$ or almost exactly twenty per cent.					
Reciprocals			x		
The reciprocal of 1 is 1/1, of 2 is 1/2, of 1/2 is 2/1, of 1/4 is 4/1, and so forth.					
Relative Numbers				x	x
The prices of coal were: in 1913, \$1.27; in 1914, \$1.17; in 1915, \$1.04. Calculating the percentages: $\frac{1.27}{1.27} = 1$, or 100%; $\frac{1.17}{1.27} = .92$, or 92%; $\frac{1.04}{1.27} = .82$, or 82%.					

TABLE XXXII (Concluded)

Computation and Application	Textbook A	B	C	D	E
If 100 stands for the price of coal in 1913, the price in 1914 relative to it will be 92, and the 1915 price, 82.					
Selling Price			x		
Cost, 50¢ Profit, 100 per cent on 50¢ = 50¢ Selling price: Cost 50¢ + Profit 50¢ = $1.00					
Taxation	x		x		
Tax rate					
On this 50 million valuation the town must raise $1,000,000 in taxes; $1,000,000 ÷ $50,000,000 = $0.02 (2¢) which is the tax rate on each $1 of property value in the town.					
The amount					
Property value respectively at $1,000,000, $2,000, and $500. A one per cent proportionate rate would produce $10,000, $20, and $5 in taxation.					
Wages		x			
If the rate is $5 per day and the laborer has worked 4½ days, he will find in his pay envelope $22.50.					
Yield or Net Return		x			
If a stock sells at $200 and pays 8% dividends annually, the ratio of the dividend, $8, to the total cost, $200, is .04, or in other words, the net return, or yield, is 4%.					

graphs, a number of calculations are required also. A set of typical exercises involving calculations follows.

1. Arrange the following items in the form of a statement of the assets and liabilities of a corporation: land and buildings, $90,000; capital stock, $100,000; bonds outstanding, $50,000; machinery and equipment, $75,000; Liberty bonds, $5,000; cash, $7,000; surplus, $60,000; taxes due, $3,260; accounts receivable, $12,000; raw materials on hand, $26,000; finished goods on hand, $20,000; accounts payable, $15,000; undivided profits, $6,740. There are 1,000 shares of capital stock outstanding. What is the par value of each share? the book value?
2. Assume a capitalization rate of 5 per cent. What is the capital value of a house yielding $500 a year net rent? of a bus franchise yielding $6,000 a year net? of a cow yielding 4 quarts of milk a day net, worth at the farm 10¢ a quart?

TABLE XXXIII

MATHEMATICAL ACTIVITIES REQUIRED IN THE ELEMENTARY ECONOMICS TEXTBOOKS EXAMINED

Activity	Textbook A	B	C	D	E	Total
Prepare a balance sheet					1	1
Prepare a bill of exchange					1	1
Prepare a budget		1				1
Prepare calculations						
Book value					3	3
Capital value			2			2
Comparative cost			2			2
Future worth					3	3
Going value			1			1
Import duty			3			3
Index number		1	3			4
Market value		1				1
Par value					1	1
Par of exchange					5	5
Percentage			1			1
Present value					3	3
Rate of yield on investment			1			1
Ratio		1	1		2	4
Tax		2	2			4
Prepare formulas					2	2
Prepare graphs						
Cost					1	1
Demand and supply curves		3	1		19	23
Diminishing utility diagram		1				1
Ordinate and abscissa of a point on a curve					1	1
Price curve		1	3		3	7
Trade diagram			1			1
Trend line			1			1
Vertical demand curve					1	1
Prepare schedules						
Demand and supply			2		2	4
Market values			1			1
Prepare statements		1	1		6	8
Interpret graphs						
Abscissa for a point on a curve					1	1
Abscissa for the point of intersection of two curves					4	4
Extremities of a curve					2	2
Ordinate for the point of intersection of two curves					4	4
Slope					2	2
Interpret tables		4				4
Total		16	26		67	109

3. If your pocketknife cost 50¢ to make and your fountain pen, $1, figure the comparative cost of each in terms of the other.
4. Suppose $100 to be deposited in a savings bank which pays four per cent interest on such deposits.
   (*a*) What would this deposit be worth in three years time if no money were withdrawn in the meantime?
   (*b*) What would it be worth in six years?
   (*c*) What would it be worth in fifteen years?
   Note: Solve (a) exactly. For (b) and (c) work out the formula only.
5. Assuming a capitalism rate of 5 per cent, what is the going value of a public utility with average annual net earnings of $100,000?
6. The last paragraph of Schedule 6 on "Tobacco and Manufactures of" in the 1930 Tariff law reads:
   "Par. 605. Cigars, cigarettes, cheroots of all kinds, $4.50 per pound and 25 per cent ad valorem, and paper cigars and cigarettes, including wrappers, shall be subject to the same duties as are herein imposed upon cigars."
   Calculate the import duty on an importation of cigars weighing 890 pounds, and billed as worth $712. On an importation of better cigars weighing the same, but billed at $2,990.
7. Suppose you are using the progressive principle and have rates as follows:
   On incomes under $2,000, tax rate is 1%
   On incomes from $2,001 to $5,000, tax rate is 3%
   On incomes from $5,001 to $10,000, tax rate is 6%
   On incomes from $10,001 to $20,000, tax rate is 10%
   On incomes above $20,000, tax rate is 15%.
   Calculate the income taxes three men with yearly incomes of $2,500, $12,000, and $50,000, respectively, must pay, (a) without an exemption, (b) with a $1,000 exemption.
8. From the following supposed schedule of prices in 1920 and 1930 prepare a simple relative index number to measure the change in the price level:

Commodity	Price in	
	1920	1930
Wheat, per bushel	$ 2.00	$ 1.00
Coal, per ton	18.00	15.00
Paper, per pound	.08	.06
Wool, per pound	.25	.30
Steel, per ton	40.00	30.00

9. The gold franc contains 4.4803 grains of fine gold. Determine the mint par of exchanges:
   (*a*) in terms of cents per franc.
   (*b*) in terms of francs per dollar.
   (*c*) in terms of francs per pound sterling.

10. What percentages of the total price of the cigars does the duty in each case represent? Refer to (6).
11. Suppose the interest rate to be five per cent. What is the present value of a promissory note for $1,000 due in two years?
12. An estate with a net rental of $100 a year sells for $2,000. What rate of yield on the investment does this give?
13. Silver has recently been selling for 30¢ an ounce; at this rate, what is the ratio between silver and gold at $20.67 an ounce?
14. What is the surtax on an income of $92,000?

TABLE XXXIV

FREQUENCY OF OCCURRENCE OF TABLES AND GRAPHS IN THE ELEMENTARY ECONOMICS TEXTBOOKS EXAMINED

Tables and Graphs	Textbook A	B	C	D	E	Total
Tables	8	76	37	59	60	240
Graphs	106	59	38	20	32	255
Broken line	30	23	17	3	6	79
Circle	38	5				43
Curved line	2	2				4
Diagram	6	10	11	4		31
Function	7		7	13	26	53
Horizontal bar	2	8	2			12
Vertical bar	21	5				26
Combination of two or more of those already listed		6	1			7

The graphs used in economics are characterized not only by their increased frequency of occurrence but also by the increased complexity of the statistical graphs and by the use of the function graph and the terms associated with its use. The classification of the types of graphs used and the frequencies of their occurrence are given in Table XXXIV. Table XXXV shows the purposes for which each type of graph was used, the items scaled, the nature of the backgrounds and rulings, and a summary of the complexity for statistical graphs. A few illustrations preceding the table will serve to show some of the types of graphs in use in economics. Figures 38-41 show more clearly some uses of the broken line graph; Figure 42, the circle graph; Figure 43, the curved line graph; Figures 44-49, the horizontal bar graph; Figures 50-53, the vertical bar graph; and Figures 54-59, some more complex statistical graphs. Figures 39 and 40 also illustrate the trend lines on statistical graphs.

These graphs have been reproduced from the following publications with the permission of the publishers:

ELY, RICHARD T., and OTHERS. *Outlines of Economics*. The Macmillan Company, 1931. By permission of The Macmillan Company.

FAUBEL, ARTHUR L. *Principles of Economics*. Harcourt, Brace and Company, Inc., 1932. By permission of Harcourt, Brace and Company, Inc.

FAY, CHARLES R. *Elements of Economics*. The Macmillan Company, 1932. By permission of The Macmillan Company.

THOMPSON, CHARLES M. *High School Economics*. Benj. H. Sanborn & Co., 1932. By permission of Benj. H. Sanborn & Co.

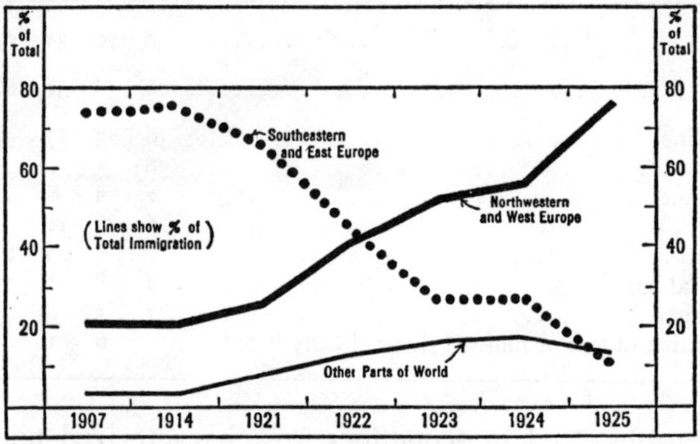

FIGURE 38. The Shifting Trend of Immigration, 1907-1925
*From Fay.**

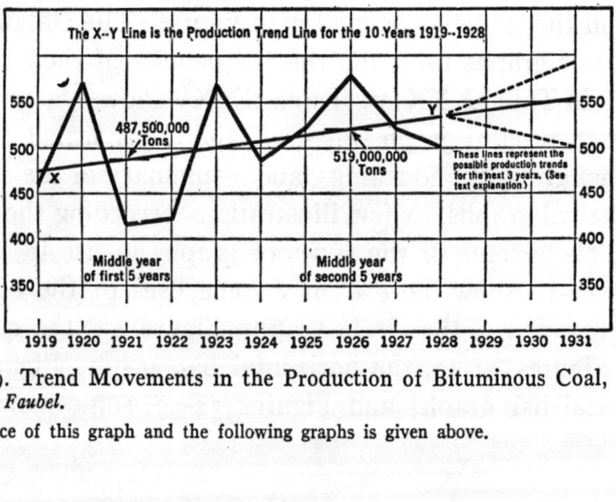

FIGURE 39. Trend Movements in the Production of Bituminous Coal, 1919-1928
From Faubel.

* The source of this graph and the following graphs is given above.

FIGURE 40. Interest Rates on 60-90 Day Commercial Paper, 1903-1916
From Ely.

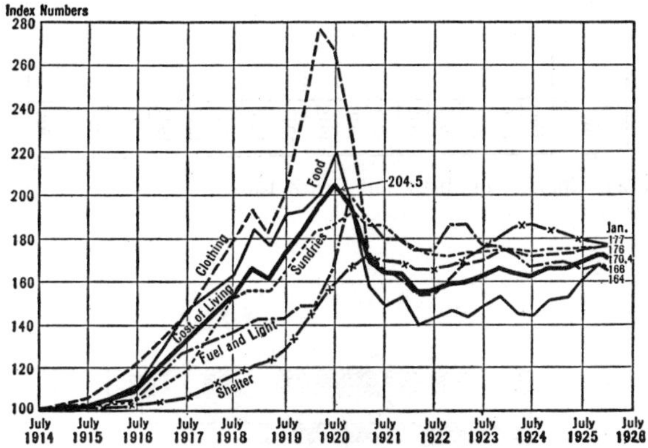

FIGURE 41. Changes in Major Items of Cost of Living
From Fay.

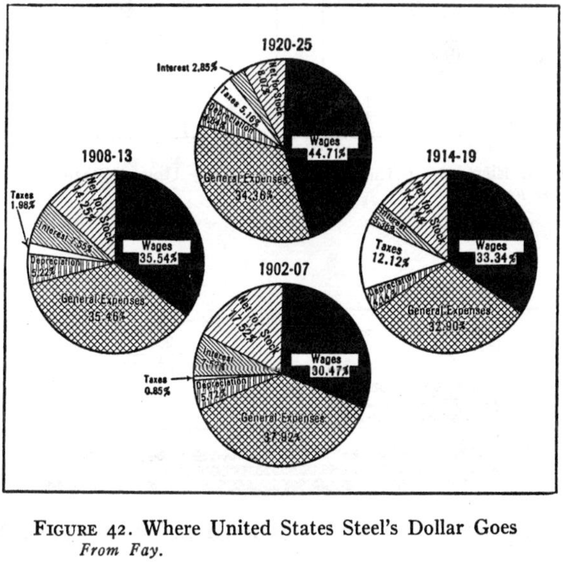

FIGURE 42. Where United States Steel's Dollar Goes
From Fay.

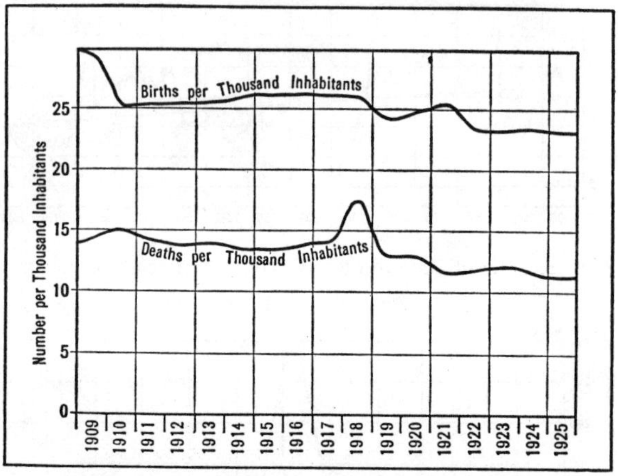

FIGURE 43. Changes in Birth and Death Rates in Continental United States
From Fay.

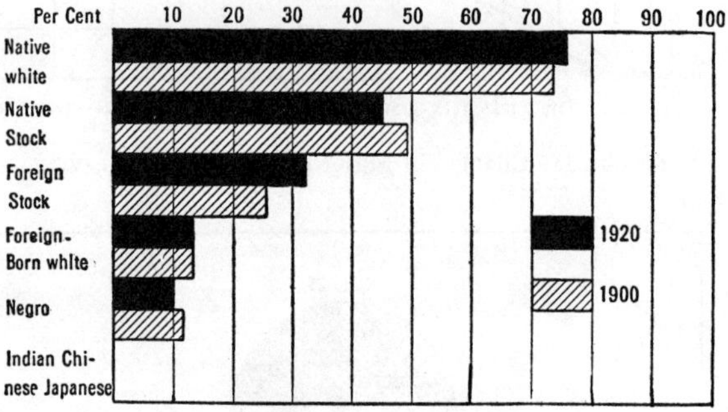

FIGURE 44. Elements in the Population of the United States, 1900 and 1920
From Fay.

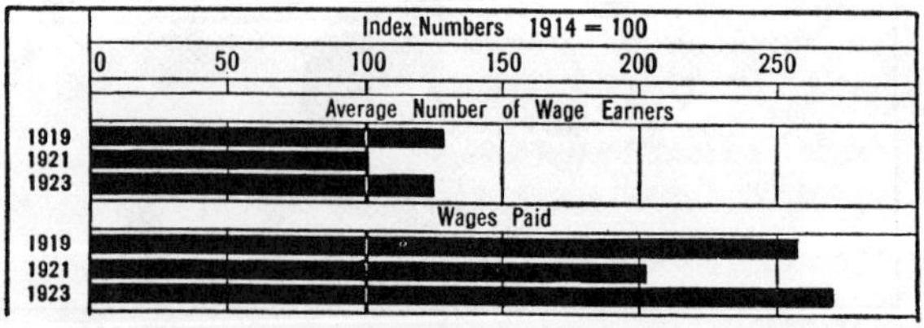

FIGURE 45. Indexes of Census Statistics of the Manufacturing Industry
From Fay.

"In statistical measurements of economic movements, it is necessary first to distinguish the secular trend or long-time movement which marks such phenomena as gold production or canal traffic."[14] The determination of a trend line to show the long-time movement or trend of bituminous coal production was described in one of the economics textbooks examined. The resulting line is shown as the trend line in Figure 39.

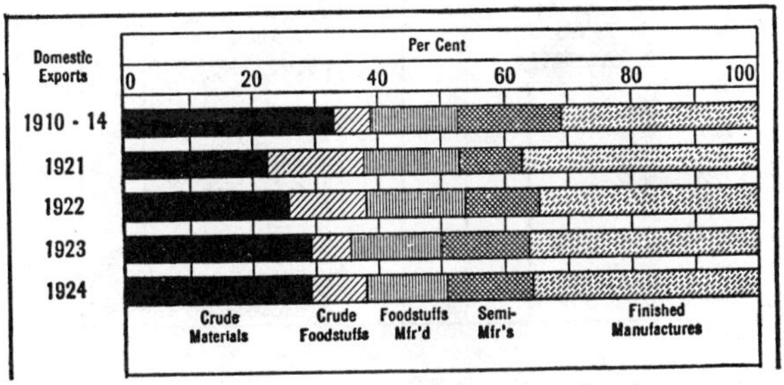

FIGURE 46. Percentage Distribution of Foreign Trade of the United States by Great Groups

From Fay.

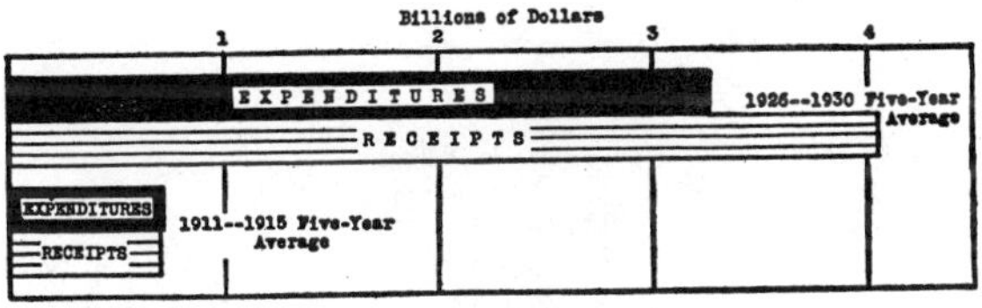

FIGURE 47
From Faubel.

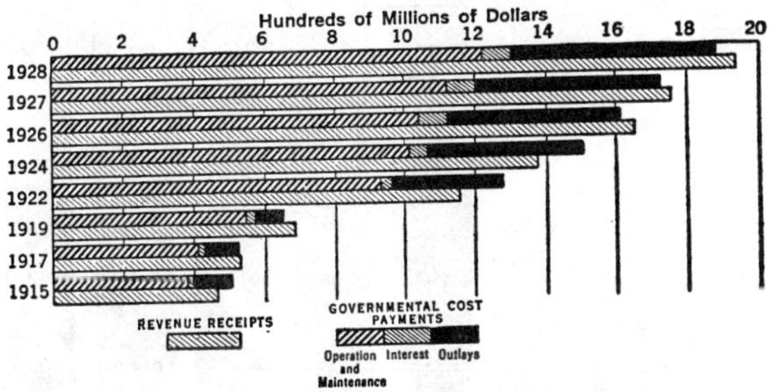

FIGURE 48
From Faubel.

[14] Ely and Others, *op. cit.*, p. 325.

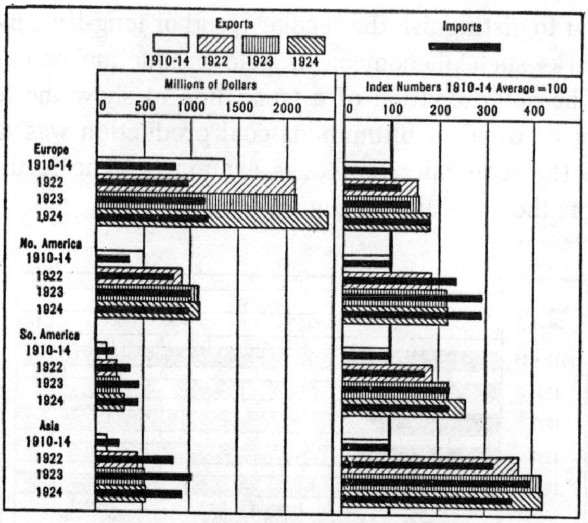

FIGURE 49. Merchandise Exports from and Imports into the United States by Continents

From Fay.

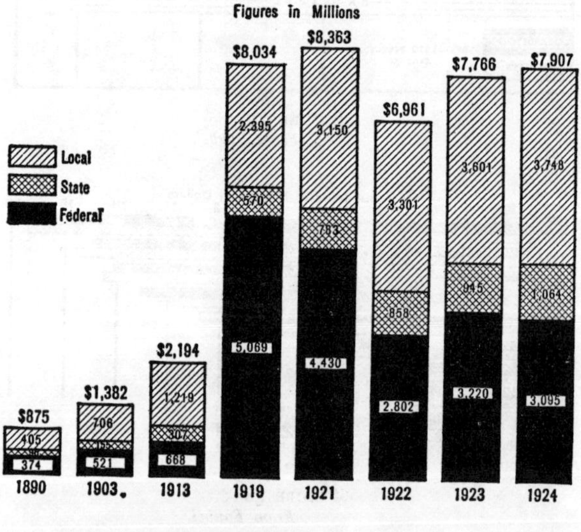

FIGURE 50. Taxation in the United States

From Fay.

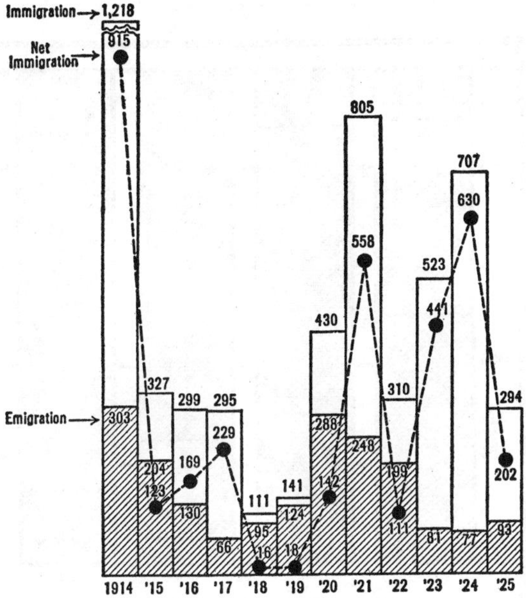

FIGURE 51. Immigration, Emigration, and Net Immigration (in thousands)
From Fay.

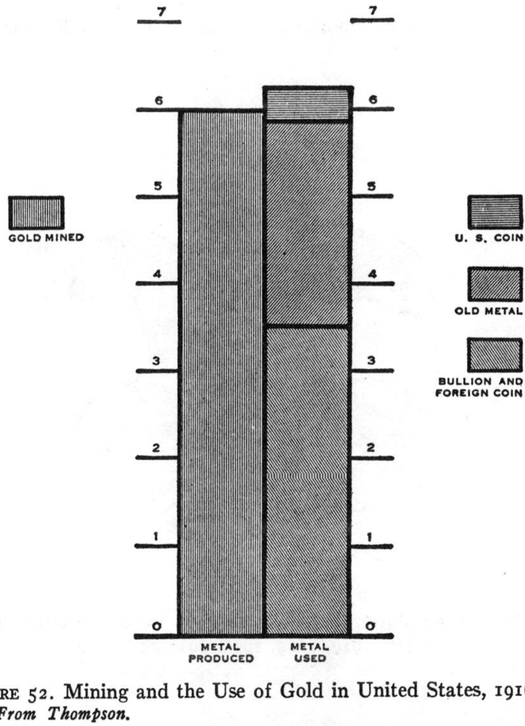

FIGURE 52. Mining and the Use of Gold in United States, 1916-1925
From Thompson.

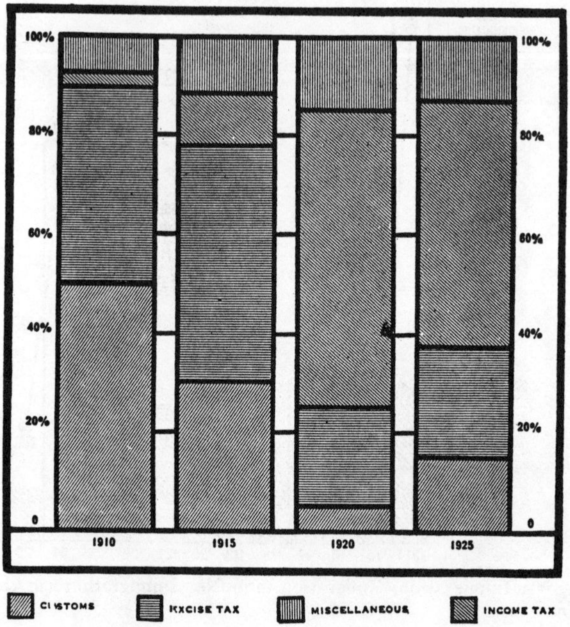

FIGURE 53. Comparative Importance of Chief Sources of Revenue, 1910-1925
From Thompson.

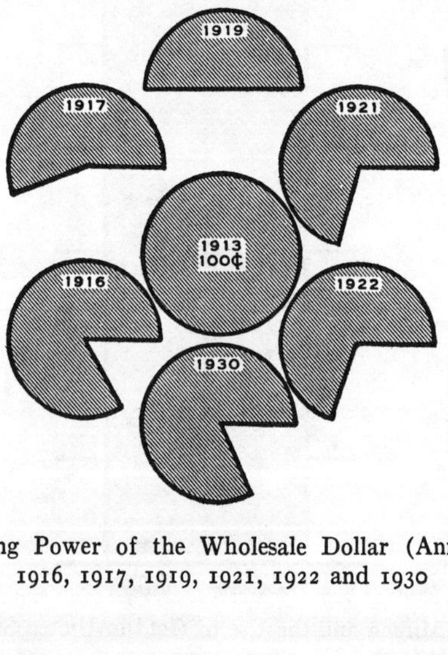

FIGURE 54. Purchasing Power of the Wholesale Dollar (Annual Average) 1913, 1916, 1917, 1919, 1921, 1922 and 1930
From Fay.

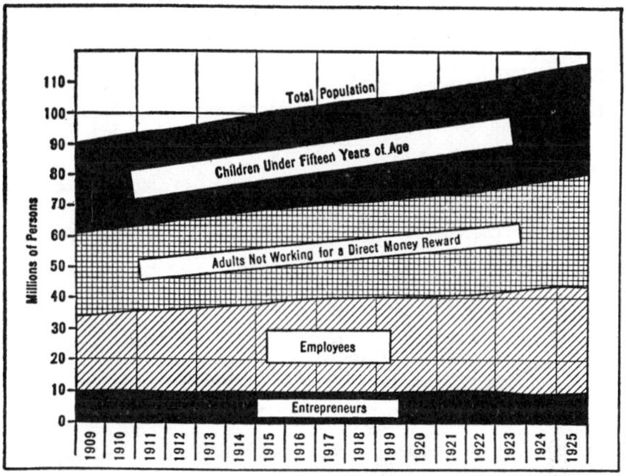

FIGURE 55. Population of the Continental United States Classified According to Occupational Status

From Fay.

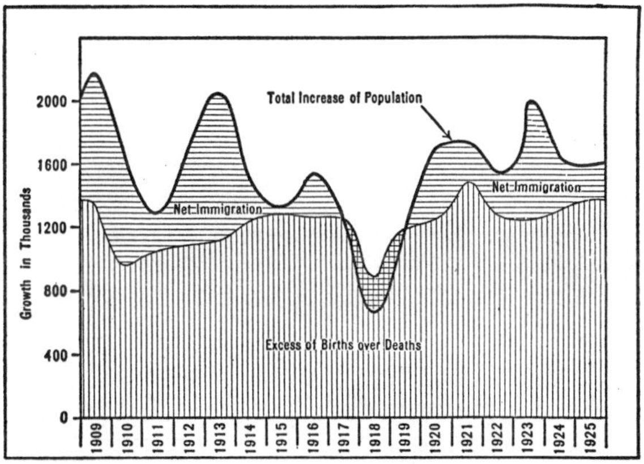

FIGURE 56. Growth in Population in Continental United States

From Fay.

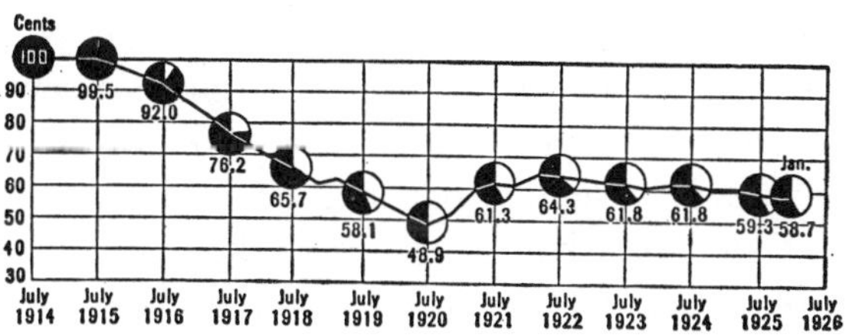

FIGURE 57. Purchasing Value of the Dollar, Based on the Cost of Living

From Fay.

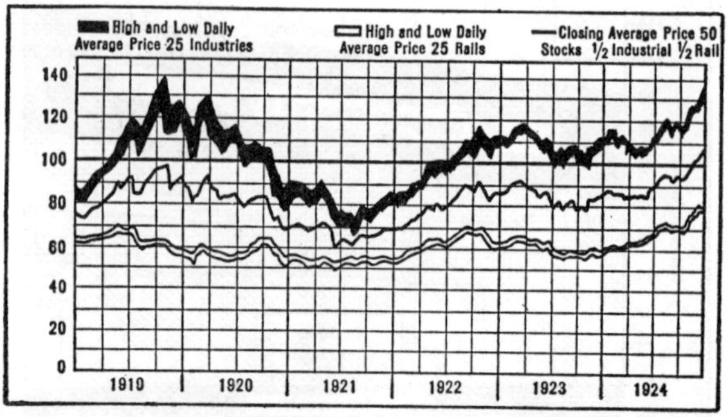

FIGURE 58. Stock Prices on New York Stock Exchange, 1910-1924
From Fay.

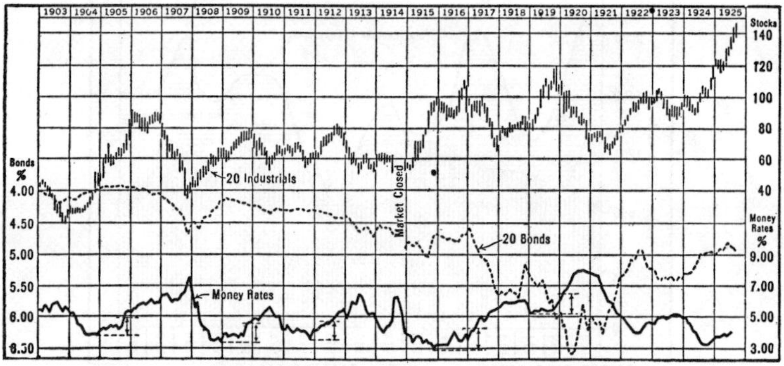

FIGURE 59. Money Rates, Bonds, and Stocks, 1903-1925
From Fay.

Table XXXVI (p. 105) shows the uses of function graphs and diagrams in the economics materials examined. Function graphs have most general use in the discussion of utility and in showing relationships among supply, demand, and price. The table also shows the terms used with functional graphs and diagrams and the numerous ways in which areas, lines, and points are named. Figures 60-70 show a few of the diagrams and graphs that occurred.

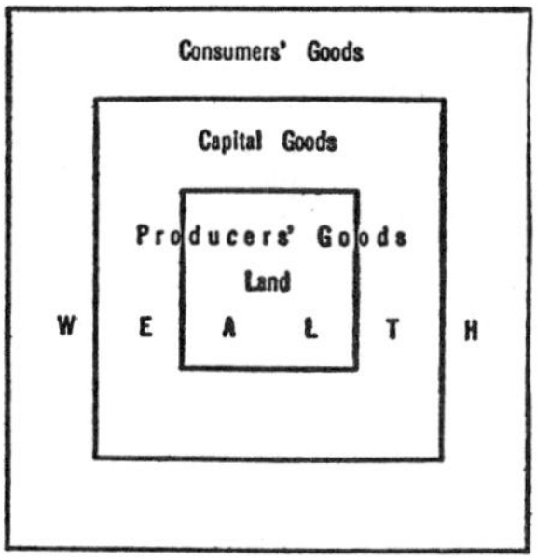

FIGURE 60. Wealth or Economic Goods
From Fay.

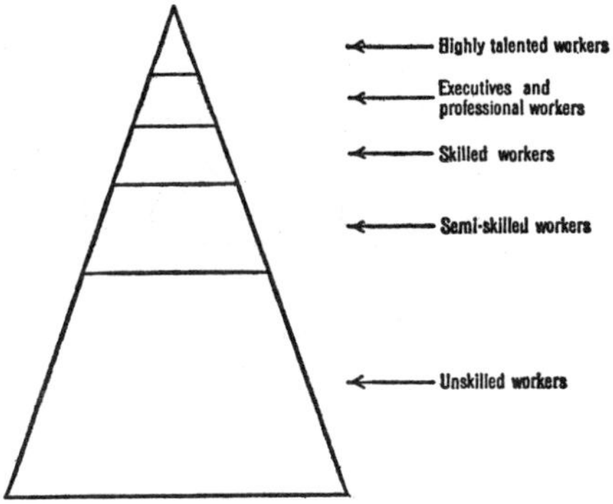

FIGURE 61. Non-Competing Groups of Laborers
From Fay.

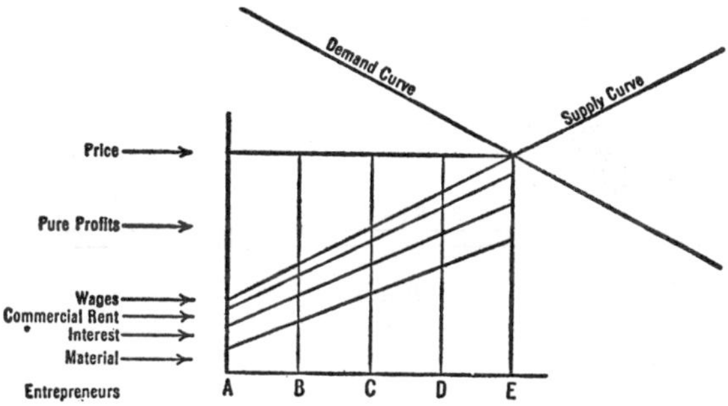

FIGURE 62. Profits Not a Part of Price
From Fay.

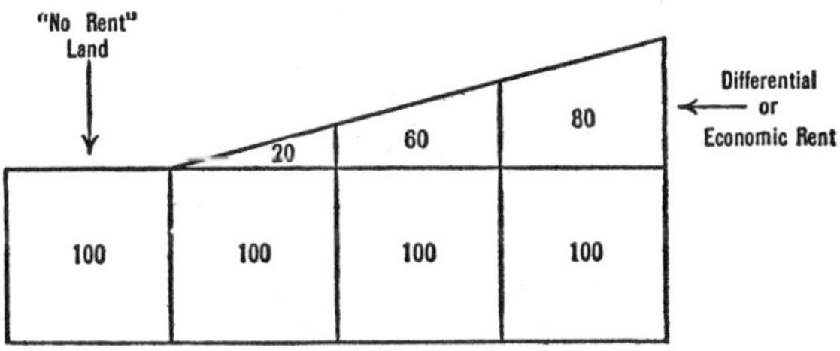

FIGURE 63. Differential or Economic Rent
From Fay.

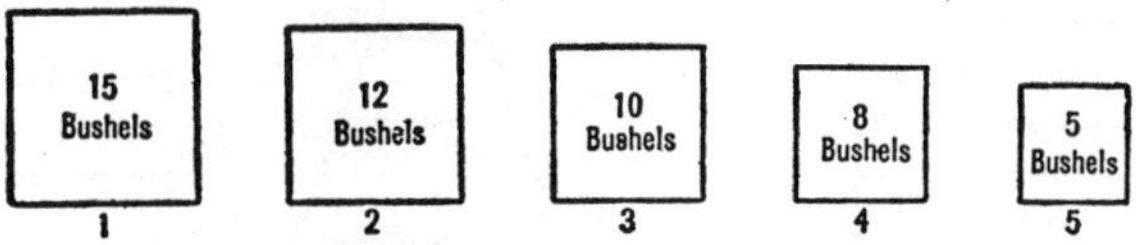

FIGURE 64. Production on Unequally Fertile Fields
From Fay.

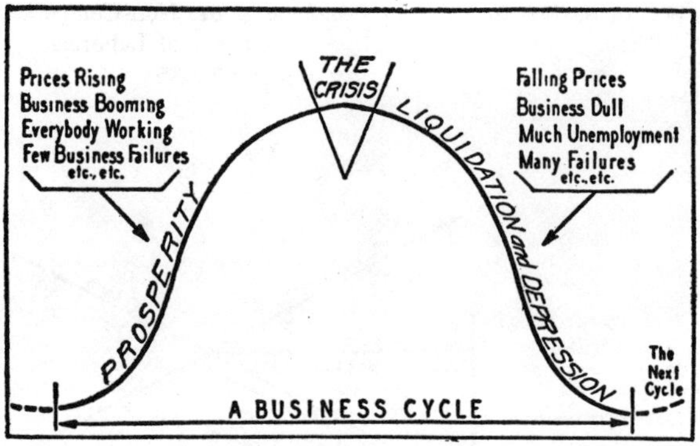

FIGURE 65
From Faubel.

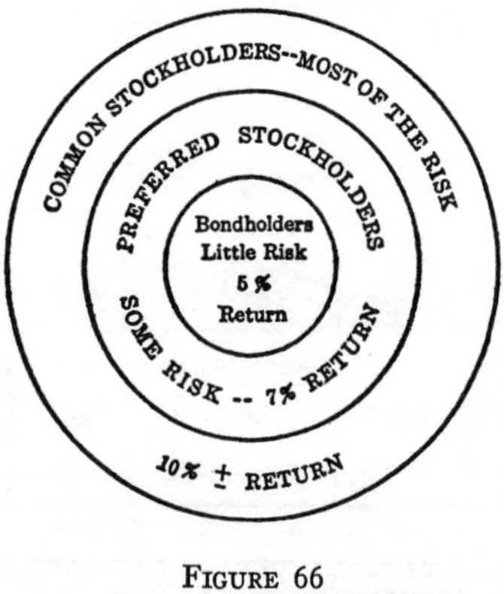

FIGURE 66
From Faubel.

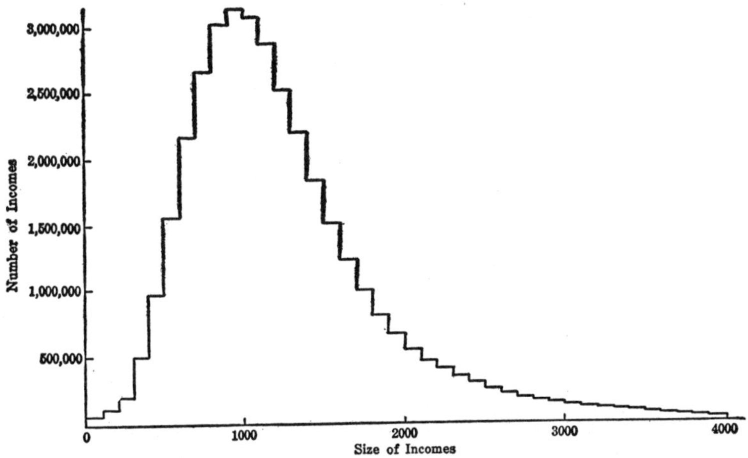

FIGURE 67. Frequency Distribution of Incomes in the United States, 1918
From Ely.

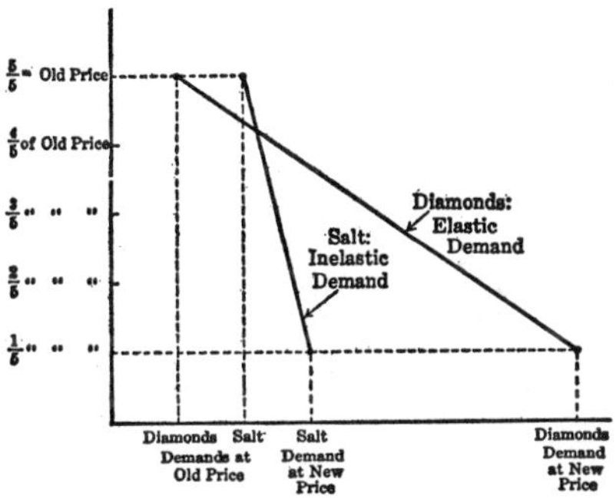

FIGURE 68
From Faubel.

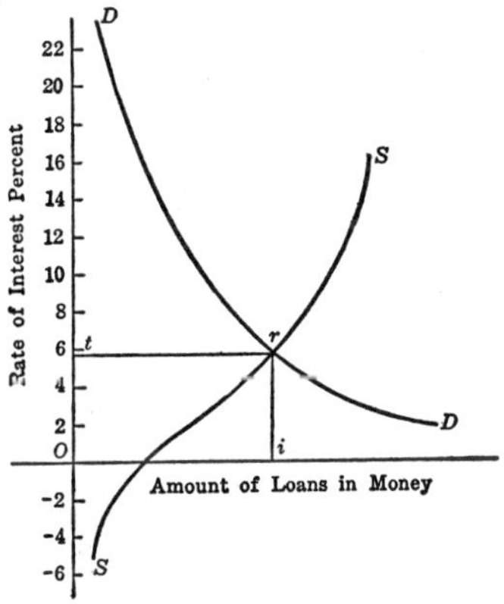

FIGURE 69
From Ely.

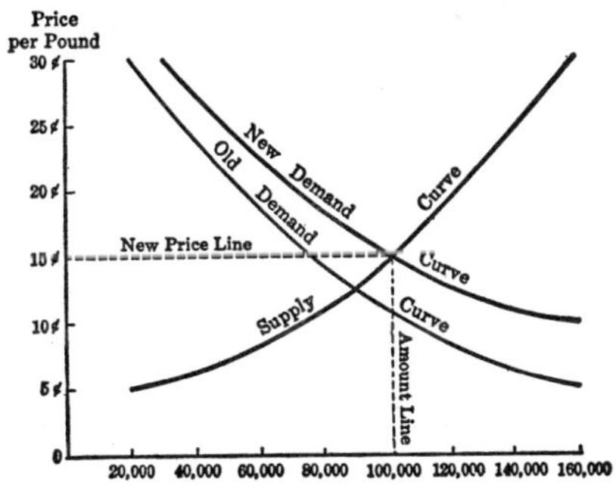

FIGURE 70
From Faubel.

TABLE XXXV

CHARACTERISTICS OF THE STATISTICAL GRAPHS USED IN THE ELEMENTARY ECONOMICS TEXTBOOKS EXAMINED

Characteristic	Frequency
Broken Line Graphs	79
Purpose—To compare or to show growth of	
Distribution (e.g., population)	1
Number (e.g., immigrants, votes, etc.)	9
Production (e.g., iron, coal, etc.)	14
Size (e.g., prices, trade, etc.)	42
Value (e.g., farms, insurance, etc.)	13
Items scaled	
Vertically—acres, automobiles, barrels, British thermal units, business failures, cents, dollars, immigrants, index numbers, members, miles, ounces, patents, people, per cents, points, pounds, tons, votes, years	79
Horizontally—dollars, months, pounds, years	79
Position of items scaled	
On left	77
On right	51
On left and on right	49
On bottom	75
On top	7
On bottom and on top	3
Rulings	
Equal rectangles	54
Equal squares	21
Unequal rectangles	4
Some subdivisions	10
Number of broken lines	
One	43
Two	24
Three	7
Five, six, or seven	5
Circle Graphs	43
Purpose—To show	
Distribution (e.g., expenses, wealth, etc.)	38
Growth (e.g., investment, production, etc.)	5
Per cents shown on graphs	31
Number of sectors	
Less than three	6
Three	7
Four	6
Five	8
Six	9
Eight, nine, eleven, or twelve	6
Nature of sectors	
Black and shaded	1
Black, white, and shaded	4

TABLE XXXV (Continued)

Characteristic	Frequency
Shaded	13
White	23
White and shaded	1
Two internally tangent circles	1
Curved Line Graphs	4
Purpose—To show	
Absorption of capital	1
Business cycle	1
Changes in birth and death rates	1
Variation in freight carried	1
Items scaled	
Vertically—dollars, inhabitants, per cents, tons	4
Horizontally—years	4
Rulings	
Equal rectangles	3
Equal squares	1
Number of lines	
One	3
Two	1
Horizontal Bar Graphs	12
Purpose—To show	
Comparison (e.g., wage earning groups, hand and machine labor, etc.)	10
Growth (e.g., power, population)	2
Items scaled	
Vertically—elements, groups, states, years	12
Horizontally—per cents, wage earners, people, dollars, hours, index numbers, tons	12
Position of items scaled	
On left	8
On bottom	3
On top	10
On bottom and at top	3
Numbers at ends of bars	1
Quantities written on bars	1
Rulings	
Vertical lines equally spaced	11
No lines	1
Number of bars	
Two, three, or four	3
Ten, sixteen, or eighteen	5
Twenty-one, thirty, sixty-four, or ninety-six	4
Nature of bars	
Black	2
Black and shaded	4
Black, white, and shaded	2
Black and white	1

TABLE XXXV (Concluded)

Characteristic	Frequency
Shaded	2
Shaded and white	1
Divided	4
Overlapping	1
Grouped in two's	6
Grouped in three's	1
Grouped in four's	1
Vertical Bar Graphs	26
Purpose—To show	
Comparison (e.g., prices, values, etc.)	22
Growth (e.g., expenditures, production, etc.)	4
Items scaled	
Vertically—bank suspensions, cars, cents, dollars, employees, establishments, index numbers, people, per cents, stockholders	26
Horizontally—sections, years, cities, metals, countries, causes	26
Position of items scaled	
On left	23
On right	22
On left and on right	21
On bottom	16
On top	1
Numbers at ends of bars	2
Numbers or quantities written on bars	11
Rulings	
Horizontal lines equally spaced	24
No lines	2
Number of bars	
Two, four, or six	8
Seven, eight, or nine	12
Twelve, fourteen, fifteen, twenty-four, or twenty-seven	6
Nature of bars	
Black	3
Black and shaded	1
Black, white, and shaded	1
Shaded	15
White	6
Divided	3
Grouped in two's	2
Grouped in three's	8

TABLE XXXVI

VOCABULARY AND CONNECTIONS WITH DIAGRAMS AND FUNCTION GRAPHS IN THE ELEMENTARY ECONOMICS TEXTBOOKS EXAMINED

Connection and Vocabulary	Textbook A	B	C	D	E	Total
Connection with						
Business cycle			x			
Demand	x	x	x	x	x	
Diminishing returns		x				
Expenses or cost				x	x	
Frequency distribution				x		
Interest		x		x		
Price	x	x	x	x	x	
Production		x	x			
Profit		x				
Rent	x	x		x		
Risks			x			
Supply	x	x	x	x	x	
Utility	x	x	x	x	x	
Wages		x		x		
Vocabulary						
Abscissa(s)				1	4	5
Abscissa of a point					1	1
Area				1		1
Area *ABC*				4		4
Area *ABCD*				1		1
Area *Abcd*				2	2	4
Area *Abcde*					1	1
Area *abcd*				3	6	9
Area above the line *lm*				2		2
Ascend					4	4
Asymmetrical				1		1
Asymmetry				1		1
Axis (axes)					7	7
Axis *LM*				1		1
Axis of abscissas					3	3
Axis of ordinates					3	3
Base axis					1	1
Base line				1	1	2
Cd (to name a curve)					18	18
Connect the points				1		1
Continuous curve					1	1
Cost curve					3	3
Cost curve *cde*					1	1
Curve(s)	3		2	6	53	64
Curve *CD*	5			4		9
Curve *cd*					19	19
Curve *CDE*				3		3
Curve *cde*					1	1

TABLE XXXVI (Continued)

Connection and Vocabulary	Textbook A	B	C	D	E	Total
Curved line				1	1	2
Curved line *lm*				1		1
Curves intersect at the point *P*					2	2
Demand curve	2	1	5	15	14	37
Deviate				2		2
Difference between *lm* and *ln*					2	2
Difference between the area *ABC* and the shaded area				1		1
Difference is *LM*, *NM*—*NL*	1					1
Direction	1				1	2
Distance					9	9
Distances measured from *P*				1		1
Distances measured vertically from line *LM*				1		1
End(s)					2	2
Expense curve				1		1
Graduated					1	1
Height				1	1	2
Height of *P*	1					1
Height *LM*	2					2
Horizontal					2	2
Horizontal base line				1		1
Horizontal cost curve					2	2
Horizontal line	1			1	7	9
Horizontal perpendicular *lm*					1	1
Imaginary vertical lines				1		1
Infinite change					1	1
Infinitesimal change					1	1
Intersect					3	3
Intersecting					1	1
Intersection					3	3
Intersection of the curves at *P*					2	2
Irregular area *ABC*				1		1
Irregular curve *CD*				1		1
L (to name a line)	5					5
l (to name a line)					2	2
LM (to name a line)	39			46		85
Lm (to name a line)				1	26	27
lm (to name a line)				3	60	63
Line					7	7
Line *LM*				5		5
Line *lm*				5	7	12
Line intersects the curve at point *P*					1	1
Mark off portions of the base line				1		1
Measured along *LM*				2		2
Measured upon the horizontal scale					1	1
Normal curve				1		1
Ordinate(s)				1	9	10
Ordinate of the point					1	1
Origin					4	4

TABLE XXXVI (Concluded)

Connection and Vocabulary	Textbook A	B	C	D	E	Total
P (to name a point)				5	3	8
p (to name a point)					11	11
Parallel					3	3
Perpendicular			1		3	4
Plot			1			1
Plotting				1	4	5
Point(s)				5	34	39
Point *p*	6				14	20
Point of intersection					3	3
Point of intersection of the two lines					1	1
Point where the two curves cross				1		1
Product of the ordinate and abscissa					1	1
Production curve				2		2
RS (to name a rectangle)				5		5
Rstu (to name a rectangle)				6		6
Rectangle(s)	1	1		4		6
Rectangle *R*	7					7
Rectangle *RS*	1			2		3
Rectangle *RSTU*				7		7
Rectangle *l*			5			5
Rectangular hyperbola					1	1
Relation between magnitudes					1	1
Right angles					1	1
Scale divisions					2	2
Skewed				1		1
Skewness				1		1
Slope(s)	1			5	4	10
Smooth curve				1	2	3
Smoothing of the curve					1	1
Straight horizontal line LM				1		1
Straight line				1	1	2
Straight vertical line					1	1
Supply curve			4	2	9	15
Supply curve *ST*				4	1	5
Symmetrical curve				2		2
Vertical demand curve					1	1
Vertical distance				1		1
Vertical distance from *P*	1					1
Vertical line *L*	3					3
Vertical ordinate					1	1
Vertical perpendicular *lm*					1	1
Vertical scale					1	1
Vertical supply curve					1	1
x axis					3	3
y axis					4	4
Zero position					2	2
Total	80	2	18	183	418	701

SUMMARY

The results of an examination of the textbooks in elementary economics show that the mathematical demands are considerably more extensive than in the more general social studies materials.

The numbers in use and many of the mathematical terms that occur in economics are similar to those that occur in the more general materials. But the frequent use of index numbers, formulas and equations, variation, and function graphs characterize significant additional uses of mathematics in economics. The statistical graphs are somewhat more complex and the computations illustrated and required and the topics with which the graphs and computations are associated are somewhat different.

The desirability of considering the needs and the connections in economics in the treatment of formulas and equations, and direct and inverse variation in the mathematics of the ninth year deserves consideration. A more extended treatment of averages, the different types and their applications; the determination of relative numbers and index numbers and a study of their uses and importance; and the connections with which function graphs are used and the types of function graphs in economics require consideration as further materials to be added to the early mathematics courses to assure provision for the training in mathematics essential to meet the demands of elementary economics.

Not all pupils study economics in high schools, and not all college students elect economics in college. But many who do not study economics in college or in high school choose books in economics for their general reading. Knowledge of the mathematics involved should be a part of the equipment for reading or studying elementary economics materials.

CHAPTER VI

CONCLUSIONS AND RECOMMENDATIONS

LIMITATIONS OF THE STUDY

THIS study has been limited to the tabulation of the mathematical concepts and processes and their social studies connections in a selected list of textbooks in the social studies adapted for the three-year period of the junior high school, for a one-year introductory course prescribed for college freshmen, and for elective courses in elementary economics at the high school and college levels. The reference materials are believed to be representative of the materials of instruction in use in the respective grades for which they are adapted, but there is considerable probability that other materials would tend to introduce some additional concepts or make less use of some of those that have been emphasized.

The study makes no pretense of having determined the entire range of the mathematics used in the grades studied or of the potential possibilities for the use of mathematics in clarifying or in emphasizing certain materials of the social studies in these grades. It has given no consideration to the extent to which habits of behavior or of thought that represent valuable outcomes of the study of mathematics may be applicable to the study of the social sciences. It has been concerned solely with concepts and processes that appear on the printed page, their frequency of occurrence at various grade levels and in various divisions of the social studies, their social studies connections, and the provision made in textbooks in mathematics for learning them.

CONCLUSIONS

All the mathematics encountered in the social studies materials examined can be thought of as occurring in some one of the categories of whole numbers, fractions, decimals, units of measure, word concepts, tables, graphs, mathematical activities, signed numbers, symbolism, mathematical facts, mathematicians, formulas and equations, computations, statements, index numbers, and variation.

The extensive use of number in the social studies testifies to the quantitative nature of the social studies materials. Number in different forms is used with a large variety of connections. Ordinal and cardinal numbers appear expressed in Arabic notation and in words. Ordinal numbers are also expressed in Roman notation and cardinal numbers in a combination of figures and words. A vast majority of the fractions are expressed in words. Ordinal numbers range in size from one-place numbers to five-place numbers, and cardinal numbers range from zero to billions in the junior high school social studies materials, and from zero to quintillions in the introductory social studies materials at the college level. Frequent use is made of the larger numbers in all the social studies materials.

The fractions most commonly used are those with denominators two, three, four, five, six, and ten. Considerably more than half are unit fractions. Fractions with large denominators are somewhat more frequent at the higher grade levels. The decimals in the social studies for the junior high school years are confined to tenths, hundredths, and thousandths and are used chiefly to express averages, per cents, and dollars and cents. Approximate ratios of exchange in economics are frequently expressed to four, five, or six decimal places. Approximate numbers are often indicated, too, by such qualifying words as about, more than, over, nearly, less than, and similar terms.

If emphasis in learning the concepts of number in arithmetic and in junior high school mathematics is placed upon the meanings of the concepts and, as in the case of Roman numerals, upon their formation, it appears safe to conclude from a study of the content of several series of arithmetic textbooks and junior high school textbooks in mathematics that the number needs of the social studies are met in these textbooks.

A large number of units of measure appear in the social studies. The units of time, length, area, and weight that are used in the normal experiences of most individuals occur with the largest frequencies. Practically no metric units appear in the junior high school social studies materials, but they do occur in the general social studies materials at the higher grade levels and in economics. Not all the units in the social studies materials appear in the mathematics textbooks, and there seems to be no justification for including them.

Ninety-six mathematical terms, which are selected from a much larger list as described on page 25 and not including the units of measure or the terms associated with measurement, are in use in the junior high school social studies materials. The most common are area, average, estimate, graph, line, per cent, proportion, rate, scale, and terms in use with the fundamental processes of arithmetic. In the more general materials at the higher grade levels a large number of terms are added in the statements of mathematical principles, in descriptions of the nature of mathematics, in discussions of mathematics as a method of thinking, in the accounts of the relation of mathematics to science, and in accounts of the development of, and refinements in, measurement. Geometric terms, terms from higher mathematics, statistical concepts, and terms used in finance and business account for the greater part of the increase in the number of terms. In economics terms used with function graphs, variation, and in certain computations peculiar to economics account for an increase in the mathematical vocabulary in the social studies. Many of the mathematical terms that are added do not occur in the arithmetic or in the junior high school mathematics textbooks and have no place in them. The understanding of many of the terms requires the study of mathematics considerably beyond the required junior high school courses. Some few terms might receive more attention in the earlier mathematics textbooks.

The mathematical symbolism used in the social studies is not extensive. The symbols found in arithmetic are in common use in the social studies. The symbols for positive and negative, the radical sign, positive and negative numbers, fractional exponents, and exponents in decimal form occur. A number of different ways of naming points, lines, and polygons are found in connection with the use of the function graph in economics. Some of the ways are in contrast to the accepted ways in mathematics, but they appear clear in their setting.

Tables and graphs are used extensively in the social studies.

The first tables that appear are very simple, and preparation for their interpretation in mathematics textbooks is mostly incidental. Most of the mathematics textbooks illustrate or make use of tables, but few treat the topic separately. Graphs are finding a very definite place in picturing the quantitative relationships in the social studies materials. Bar graphs, circle graphs, and line graphs are used in the materials for the seventh year to show the comparison or growth of a number of different items. They occur with about the same complexity in the materials for the eighth and ninth years. They are less frequent but more complex in the general materials at the higher grade levels. The most frequent use of statistical graphs occurs in economics. Function graphs are introduced also, but are confined primarily to the economics textbooks recommended for college use. Several figures in Chapter V illustrate the type and complexity of statistical and function graphs appearing in economics. The use and construction of trend lines are discussed in economics also. The construction and interpretation of various types of graphs are required at all grade levels. Preparation for the construction and interpretation of the graphs occurring in the social studies materials of the seventh year is not made sufficiently early in many textbooks in arithmetic. Little attention is given in mathematics textbooks to the type of situation in which the function graph is used in economics.

The computations illustrated in the junior high school social studies materials are few in number, comprising in the fusion materials the method of determining the century in which a given date lies, finding interest on a loan, obtaining a tax rate, and finding the amount of tax; in the workbooks accompanying certain fusion materials, the method of calculating per cents; and in the geography materials, the method of calculating standard time and of finding latitude and longitude. Other computations illustrated in economics include finding simple and weighted averages, balances, book value, capital, capitalization, comparative cost, sum of denominate numbers, discount, dividends, par of exchange, index numbers, simple and compound interest, per cents, present worth, ratios, reciprocals, series, relative numbers, selling price, tax rate, amount of tax, amount of wages, and yield or net return on stock purchases. All the processes involved in these computations are found in the study of arithmetic. The applications and the combinations of these processes in the social studies are quite new in many cases.

The mathematical activities required of pupils in the social studies in the junior high school grades include a few simple exercises involving calculations and the construction of graphs, time lines, and drawings to scale. The mathematical activities required in economics are somewhat more numerous and include a number of computations, the construction of statistical and function graphs, and the preparation of formulas, balance sheets, budgets, bills of exchange, financial statements, and tables. The mathematical activities described in the social studies materials examined include the making of scale drawings; the experimental comparison of areas; the method of constructing bar graphs and time lines, and projections in map making in the junior high school materials; and the method of interpreting and constructing function graphs in economics.

The discussion of the part that mathematics has had in social progress includes the statements of a number of geometric principles and the names of a number of celebrated mathematicians. In the introductory social studies materials at the college level the method of reaching conclusions in geometry is shown as the method of reaching conclusions in problems not related to geometry. The method used by mathematicians in developing certain phases of mathematics is exhibited as the method the mathematicians and others use in solving problems and in attaining marked achievements in other fields. The quotations in Chapter IV illustrate some applications.

The learning involved in the understanding of the geometric facts cannot be expected to be provided for pupils below the tenth year. In mathematics textbooks mathematicians are usually mentioned, if at all, only incidentally in connection with their mathematical achievements. A parallel study of their other achievements ought to prove of value in making clearer the applicability of the method of mathematics in the solution of problems in other fields.

Practically no use is made of the equation or formula in the social studies other than in economics. Formulas for finding the amount when interest is compounded, for finding the geometric average, and for the rectangular hyperbola occur. These are formulas found in the mathematics textbooks as well. The formulas of exchange, price, production, index numbers, profits, and capital are associated with the concepts of economics and seldom occur in mathematics textbooks. The study of the meaning, development, manipulation, simplification, and substitution with the formula are emphasized in mathematics in grades seven, eight and nine. All this study is necessary to undersand all the work with formulas in all the economics textbooks examined. The extent of the knowledge required varies considerably with the economics textbook in use.

Index numbers appear in only one situation in the more general social studies materials examined at the college level. They occur in all the economics textbooks examined, including those recommended for high school use. The method of computing index numbers is explained in each economics textbook, and their use is chiefly in connection with price changes. The method and detail in which they are computed determine the simpler computations involved. The simple arithmetic average, the weighted average, and the determination of per cents frequently are included. In view of the use of index numbers and weighted averages in high school social studies materials, some attention might be given to them in mathematics courses preceding their appearance in the social studies materials.

Variation, the dependence of one thing upon another, is present in the learner's experience in his earliest years of learning and in the elementary social studies materials. It is difficult to determine when this variation should be treated as mathematical variation. Variation is considered in this study as a part of the mathematical content where the ideas of variable, constant, direct, and inverse are associated with relationships among supply, demand, price, and geometric concepts. It is mentioned also in connection with formulas and equations. This type of variation, or dependence, appears in two of the three junior high school mathematics series examined, but the connections with which variation is developed in these textbooks in mathematics differ widely from the connections with which variation occurs in economics.

The provision for learning the mathematics encountered in the social studies materials examined, as determined by considering the defined terms and the explained processes in arithmetic and junior high school mathematics textbooks, is apparently adequate for the understanding of integers, fractions, and decimals; the units of measure most frequently used; a majority of the graphs beyond the seventh year; signed numbers and the symbolism involved; tables; and a majority of the word concepts that occur. Several computations found in the exposition of the social studies and required in the activities have distinctly new uses. Several word concepts, the notion of variation, mathematical principles, index numbers, averages, function graphs, and contacts with mathematicians are not provided for adequately in mathematics textbooks adapted for use below the tenth year. There is no implication that adequate preparation can be given in all the topics demanded in the social studies in mathematics courses through the ninth grade. If all pupils are to receive the mathematical training necessary to a complete understanding of all the mathematics encountered in all the social studies materials examined, it is more probable that mathematics courses considerably beyond the ninth year must be considered as desirable training.

The connections with which mathematical processes and concepts are used at the various grade levels and in the different divisions of the social studies are means of showing the many different uses that the social studies have discovered for mathematics in clarifying and in emphasizing certain topics. They show the quantitative nature of the social studies, and in learning they represent a means of conveying to pupils in mathematics courses the use of mathematics in situations that are assumed to be real situations and of giving to pupils an opportunity to appreciate the relations among various phases of departmentalized subject matter.

RECOMMENDATIONS

The findings of this study suggest the advisability of recognizing in the teaching of mathematics:

1. That concepts of quantity and quantitative relationships are used extensively in the materials of the social studies at all grade levels.

2. That many of the connections with which these concepts and relationships occur in the social studies can be used to vitalize the teaching of mathematics and to prepare for more intelligent work in the social studies. Many references in this volume too numerous to repeat here, show that at the junior high school level very definite uses are made of decimals, averages, rates, tables, graphs, and other mathematical concepts; and that certain formulas and equations, certain combinations of arithmetic processes and their applications, the uses of variation, and statistical and function graphs in economics can be used to advantage in mathematics textbooks.

3. That the social studies demand primarily a clear understanding of the language used in mathematics and of the meanings of mathematical concepts and processes. This conclusion recognizes the fact that, although junior high school pupils are required to perform only a very limited number of relatively simple calculations and to engage in few other mathematical activities, they are required frequently to understand the meaning of addition, sum, and other terms associated with arithmetic processes; the meaning of area, circle, hemisphere, parallel, and other geometric concepts; the processes of estimating, measuring, and varying; and the association of bar graphs, circle graphs, and line graphs with their uses. The tables in Chapter II reveal in more detail the extent of the meanings of mathematical concepts that are involved in the junior high school social studies. The language used with function graphs, with formulas and equations, with variation, and with certain computations emphasizes further the need for interpretations in terms of meanings in the study of economics.

4. That certain processes and certain concepts in mathematics require a special consideration not given in the mathematics textbooks examined to meet the needs of the social studies. The treatment of averages in elementary mathematics textbooks is ordinarily limited to the simple arithmetic average. Both the weighted average and the geometric average occur in economics. The meaning of variation as it is used in economics is not considered in the treatment of variation in elementary mathematics textbooks. The consideration of measurement as it occurs in economics is neglected throughout the whole subject of the nature and accuracy of

measurement in elementary mathematics. It is not within the province of this report to assert that these mathematical needs of the social studies and other needs listed on preceding pages can be met in the mathematics of the junior high school. Yet the situation is fairly clear. Pupils who undertake the study of economics in the eleventh or twelfth grades in the senior high school often do so with little preparation in mathematics beyond the ninth grade or, in some cases, an additional year of plane geometry in the tenth year. Unless these pupils get the necessary training in mathematics in the junior high school grades, they will probably find it necessary to acquire it in their social studies courses.

5. That extensive use of various types of statistical graphs in the social studies of the seventh year requires a consistently earlier introduction of these graphs in mathematics courses, probably in the sixth year.

6. That the social studies can profit from more attention in the teaching of mathematics to inculcating appreciation for the method of mathematics and the place of mathematics in the development of society and its potential possibilities in the development of the individual. There is abundant opportunity to begin to give attention to this phase in elementary courses in mathematics. Of utmost importance in attaining this objective is the understanding of the function concept. A thorough understanding of the function concept in mathematics must be associated with a better understanding of the realm and purpose of mathematics and a fuller appreciation of the application of the method of mathematics to the social studies and other fields. Although such understanding and appreciation are of considerable value even in the junior high school grades, the need is very evident in the social studies at the college level, where the method of mathematics and the place of mathematics in the development of society account for the frequent mention of mathematics and for the numerous mathematical principles and mathematical concepts that occur in the social studies at this level.

7. That the calculation of index numbers, the construction of trend lines in statistical graphs, and the applications of the function graph in economics receive no attention in mathematics textbooks below the tenth year. The discussion in the fourth

recommendation concerned itself with topics inadequately treated in mathematics textbooks, but the conclusions are equally valid for these topics that do not occur at all.

8. That certain terms, facts, and principles of mathematics that occur in the social studies materials at the college level ordinarily occur in mathematics courses after the ninth year. Axioms and postulates of Euclidean geometry, theorems and corollaries from plane and solid geometry, mathematical laws from physical science, and statistical, geometric, and higher mathematics concepts are included.

9. That more than incidental attention might be given to contacts furnished by mathematicians in correlating mathematics with other subjects. Interest in personalities and in achievements and contributions might be utilized in mathematics in considering the achievements and contributions of mathematicians in mathematics and in fields not usually associated with mathematics, to show the power of mathematics and its correlation with other subjects.

10. That pupils who are likely to continue the study of the social sciences beyond the early secondary period should be made aware of the training in mathematics that is desirable. Specific recommendations must consider the mathematical needs of the social studies, the courses in mathematics that are available in a given situation, and the potential and immediate desires and abilities of the individual pupil. A knowledge of Euclidean geometry and the training acquired in its study should prove of value in interpreting later materials in the social studies. Courses in general mathematics can assist further in meeting the mathematical demands of the social studies and in making pupils aware of these demands.

The extreme diversity and the wide range of the mathematics encountered in the social studies precludes the possibility of stating the recommendations with the detail that they deserve. Some attention must be given to the tables, figures, and conclusions in the previous chapters along with the conclusions and recommendations in this chapter.

Within certain defined limits this investigation has been concerned with the use of mathematics in the social studies. "Any person to be really 'educated,' should have been taught the importance of mathematics as an instrument of material conquests and of social organization, and should be able to appreciate the value and significance of an ordered system of mathematical ideas."[1]

[1] Nunn, T. Percy, *The Teaching of Algebra*, p. 17. 1919.

APPENDIX

I

JUNIOR HIGH SCHOOL SOCIAL STUDIES TEXTBOOKS EXAMINED

Seventh Year

Atwood, Wallace W. and Thomas, Helen G. *Nations Beyond the Seas, Book Three.* Ginn and Company, Boston, 1930. 260 pages.

Burnham, Smith. *Our Beginnings in Europe and America.* John C. Winston Company, Philadelphia, 1930. 254 pages.

Rugg, Harold. *Changing Civilizations in the Modern World.* Ginn and Company, Boston, 1930. 521 pages.

West, Ruth and West, Willis M. *The New World's Foundations in the Old.* Allyn and Bacon, New York, 1929. 296 pages.

Eighth Year

Rugg, Harold. *A History of American Civilization.* Ginn and Company, Boston, 1930. 624 pages.

Rugg, Harold. *A History of American Government and Culture.* Ginn and Company, Boston, 1931. 543 pages.

Rugg, Rugg, and Schweppe. *America's March Toward Democracy, Parts I and II.* Lincoln School of Teachers College, Columbia University, New York, 1924. 237 pages.

Rugg, Rugg, and Schweppe. *Explorers and Settlers Westward Bound.* Lincoln School of Teachers College, Columbia University, New York, 1924. 111 pages.

Rugg, Rugg, and Schweppe. *The Mechanical Conquest of America.* Lincoln School of Teachers College, Columbia University, New York, 1924. 35 pages.

Ninth Year

Edmonson, J. E. and Dondineau, A. *Citizenship through Problems.* The Macmillan Company, New York, 1927. 265 pages.

Hughes, R. O. *A Textbook in Citizenship.* Allyn and Bacon, New York, 1923. 406 pages.

Workbooks

Rugg, Harold and Mendenhall, James E. *Pupil's Workbook* to Accompany *Changing Civilizations in the Modern World.* Ginn and Company, Boston, 1930. 99 pages.

Rugg, Harold and Mendenhall, James E. *Pupil's Workbook* to Accompany *History of American Civilization.* Ginn and Company, Boston, 1930. 57 pages.

RUGG, HAROLD AND MENDENHALL, JAMES E. *Pupil's Workbook* to Accompany *A History of American Government and Culture.* Ginn and Company, Boston, 1931. 86 pages.

Additional Books

SMITH, J. RUSSEL. *Human Geography, Book Two.* John C. Winston Company, Philadelphia, 1930. 504 pages.

TRYON, ROLLA M. AND LINGLEY, CHARLES R. *The American People and Nation.* Ginn and Company, Boston, 1929. 654 pages.

II

ARITHMETIC TEXTBOOKS EXAMINED

ANDERSON, R. F. AND CADE, G. N. *Arithmetic for Today, Books One and Two.* Silver, Burdett and Company, New York, 1931.

BROWN, J. C., MIRICK, H. C., GUY, J. F., AND ELDREDGE, A. C. *Champion Arithmetics, Grades Three and Five.* Row, Peterson and Company, Evanston, Ill., 1932.

———. *Grade Four.* 1931.

———. *Grade Six.* 1933.

CLARK, J. R., OTIS, A. S., AND HATTON, C. *Modern-School Arithmetics, Books One and Two.* World Book Company, Yonkers-on-Hudson, New York, 1929.

DURELL, F., GILLET, H. O., AND DURELL, T. J. *The New Day Arithmetics, Elementary and Intermediate Books.* Charles E. Merrill Company, New York, 1930.

SMITH, D. E., LUSE, E. M., AND MORSS, E. L. *The Problem and Practice Arithmetics, First and Second Books.* Ginn and Company, Boston, 1929.

STONE, J. C. AND MILLS, C. N. *The Unit Mastery Arithmetic, Primary and Intermediate Books.* Benj. H. Sanborn & Co., Chicago, 1932.

STRAYER, G. D. AND UPTON, C. B. *Strayer-Upton Arithmetics, Lower and Middle Grades.* American Book Company, New York, 1929.

THORNDIKE, E. L. *The Thorndike Arithmetics, Books One and Two.* Rand, McNally and Company, Chicago, 1929.

III

JUNIOR HIGH SCHOOL MATHEMATICS TEXTBOOKS EXAMINED

BETZ, WILLIAM. *Junior Mathematics for Today, Book One.* Ginn and Company, Boston, 1933.

BETZ, WILLIAM. *Junior Mathematics for Today, Book Two.* Ginn and Company, Boston, 1933.

BETZ, WILLIAM. *Algebra for Today.* Ginn and Company, Boston, 1929.

BRESLICH, E. R. *Seventh-Year Mathematics.* The Macmillan Company, New York, 1930.

BRESLICH, E. R. *Eighth-Year Mathematics.* The Macmillan Company, New York, 1930.

BRESLICH, E. R. *Ninth-Year Mathematics.* The Macmillan Company, New York, 1930.

EDGERTON, E. I. AND CARPENTER, P. A. *A First Course in the New Mathematics.* Allyn and Bacon, New York, 1934.

Edgerton, E. I. and Carpenter, P. A. *A Second Course in the New Mathematics.* Allyn and Bacon, New York, 1934.

Edgerton, E. I. and Carpenter, P. A. *New Elementary Algebra.* Allyn and Bacon, New York, 1933.

IV

COLLEGE SOCIAL STUDIES TEXTBOOKS EXAMINED

Beard, Charles A. and Beard, Mary A. *The Rise of American Civilization, Volume II.* The Macmillan Company, New York, 1927. 800 pages.

Carman, Harry J. *Social and Economic History of the United States, Volume I.* D. C. Heath and Company, Boston, 1930. 573 pages.

Hammond, J. L. and Hammond, Barbara. *The Rise of Modern Industry.* Harcourt, Brace and Company, New York, 1926. 262 pages.

Hayes, Carlton J. H. *A Political and Social History of Modern Europe, Volume II.* The Macmillan Company, New York, 1928. 869 pages.

Langsam, Walter C. *The World Since 1914.* The Macmillan Company, New York, 1933. 657 pages.

Marvin, F. S. *The Living Past.* Oxford University Press, London, 1931. 290 pages.

Randall, John H. *The Making of the Modern Mind.* Houghton Mifflin Company, Boston, 1926. 642 pages.

Robinson, Edward S. *Man as Psychology Sees Him.* The Macmillan Company, New York, 1932. 359 pages.

Storck, John. *Man and Civilization.* Harcourt, Brace and Company, New York, 1927. 445 pages.

V

ECONOMICS TEXTBOOKS EXAMINED

High School

Faubel, Arthur L. *Principles of Economics.* Harcourt, Brace and Company, New York, 1932. 510 pages.

Fay, Charles R. *Elements of Economics.* The Macmillan Company, New York, 1932. 593 pages.

Thompson, Charles M. *High School Economics.* Benj. H. Sanborn & Co., Chicago, 1932. 498 pages.

College

Ely, Richard T. and Others. *Outlines of Economics.* The Macmillan Company, New York, 1931. 822 pages.

Fairchild, F. P., Furniss, E. L., and Buck, N. S. *Elementary Economics Volume I.* The Macmillan Company, New York, 1932. 594 pages.

BIBLIOGRAPHY

(*Textbooks are listed in an Appendix*)

BEARD, CHARLES A. *A Charter for the Social Sciences in the Schools.* Charles Scribner's Sons, New York, 1932.

BOWLEY, A. L. *The Mathematical Groundwork of Economics.* Oxford University Press, London, 1924.

COLUMBIA UNIVERSITY. *Announcement of Columbia College,* New York, April 7, 1934.

COUNTS, GEORGE S. *The Social Foundations of Education.* Charles Scribner's Sons, New York, 1934.

COXE, WARREN W. "When Is a School a Junior High School?" *Junior-Senior High School Clearing House,* V, pp. 72-79, October, 1930.

CRAGG, M. E. *The Mathematics Found in Social Science Textbooks.* Unpublished Master's Thesis. University of Chicago, Chicago, 1925.

DAWSON, EDGAR AND OTHERS. *Teaching the Social Studies.* The Macmillan Company, New York, 1927.

DAWSON, EDGAR. "The History Inquiry." *Historical Outlook,* XV, pp. 239-272, June, 1924.

DUNN, ARTHUR W. *The Social Studies in Secondary Education.* Bulletin, 1916, No. 28. Bureau of Education, Washington, D. C.

EVERETT, JOHN P. "Algebra and Mental Perspective." *Seventh Yearbook of the National Council of Teachers of Mathematics,* pp. 19-51. Teachers College, Columbia University, New York, 1932.

FANCLER, D. G. AND CRAWFORD, C. C. *Teaching the Social Studies.* University of Southern California, Los Angeles, 1932.

FEATHERSTONE, WILLIAM B. "An Integrated Curriculum in the Secondary Schools." *California Quarterly of Secondary Education,* VIII, pp. 337-348, June, 1933.

FISHER, IRVING. "The Application of Mathematics to the Social Sciences." *Sixth Yearbook of the National Council of Teachers of Mathematics,* pp. 1-17. Teachers College, Columbia University, New York, 1931.

GAMBRILL, J. MONTGOMERY. "Experimental Curriculum-Making in the Social Studies." *Historical Outlook,* XIV, pp. 384-406, December 1923, and XV, pp. 37-55, 84-89, January, 1924.

HASKELL, ALLAN C. *How to Make and Use Graphic Charts.* Codex Book Company, New York, 1919.

JUDD, CHARLES H. *Proceedings of the Forty-Eighth Annual Meeting of the Associated Academic Principals of the State of New York,* 1930.

JUDD, CHARLES H. *Psychology of Secondary School Subjects.* Ginn and Company, Boston, 1927.

KILPATRICK, WILLIAM H. *Education and the Social Crisis.* Liveright, Inc., New York, 1932.

KIMMEL, WILLIAM G. *Instruction in the Social Studies.* Bulletin, 1932, No. 17. Bureau of Education, Washington, D. C.

KOCH, E. H. AND SCHLAUCH, W. S. "Topics in High School Mathematics." *Contributions to Education*, Volume I, pp. 351-359. New York Society for the Experimental Study of Education, World Book Company, Yonkers-on-Hudson, N. Y., 1924.

LOMAX, P. S. AND TONNE, H. A. *Problems of Teaching Economics.* Prentice-Hall, Inc., New York, 1932.

NUNN, T. PERCY. *The Teaching of Algebra.* Longmans, Green and Company, New York, 1919.

OGBURN, W. F. AND GOLDENWEISER, A. *The Social Sciences and Their Interrelations.* Houghton Mifflin Company, Boston, 1927.

REEVE, WILLIAM D. "United States." *Fourth Yearbook of the National Council of Teachers of Mathematics*, pp. 131-186. Teachers College, Columbia University, New York, 1929.

REPORT FOR THE SOCIAL SCIENCE RESEARCH COUNCIL. "Collegiate Mathematics Needed in the Social Sciences." *American Mathematical Monthly*, XXXIX, pp. 569-577, December, 1932.

REPORT OF THE COMMISSION ON THE SOCIAL STUDIES. *Conclusions and Recommendations of the Commission.* Charles Scribner's Sons, New York, 1934.

REPORT BY THE NATIONAL COMMITTEE ON MATHEMATICAL REQUIREMENTS. *The Reorganization of Mathematics in Secondary Schools.* Dartmouth Press, Hanover, N. H., 1923.

RUGG, E. U. AND DEARBORN, N. H. *Social Studies in Teachers Colleges and Normal Schools.* Colorado State Teachers College, Denver, 1928.

SANFORD, VERA. *A Short History of Mathematics.* Houghton Mifflin Company, Boston, 1930.

SCHAAF, W. L. *Mathematics for Junior High School Teachers.* Johnson Publishing Company, Richmond, Va., 1931.

SHIELDS, H. G. "School and College Courses in Economics." *Journal of Educational Sociology*, IV, pp. 154-162, November, 1930.

SMITH, D. E. AND REEVE, W. D. *The Teaching of Junior High School Mathematics.* Ginn and Company, Boston, 1927.

Social Science, Junior High School, Grades Seven, Eight, and Nine. Course of Study Monograph, No. 3. Public Schools, Denver, 1931.

THORNDIKE, E. L. AND OTHERS. *The Psychology of Algebra.* The Macmillan Company, New York, 1923.

WILSON, HOWARD E. *The Fusion of the Social Studies in Junior High Schools.* Harvard University Press, Cambridge, Mass., 1933.

WILSON, H. E. AND ERB, B. P. "A Survey of Social Studies Courses in 301 Junior High Schools." *School Review*, XXXIX, pp. 497-507, September, 1931.

YOUNG, J. W. A. *The Teaching of Mathematics in the Elementary and Secondary School.* Longmans, Green and Company, New York, 1929.